T0076304

Is Math Real?

Also by Eugenia Cheng

x + y: A Mathematician's Manifesto for Rethinking Gender

The Art of Logic in an Illogical World

Beyond Infinity: An Expedition to the Outer Limits of Mathematics

How to Bake Pi: An Edible Exploration of the Mathematics of Mathematics

Is Math Real?

How Simple Questions Lead Us to Mathematics' Deepest Truths

EUGENIA CHENG

BASIC BOOKS
New York

Basic Books
Hachette Book Group
1290 Avenue of the Americas, New York, NY 10104
www.basicbooks.com

Printed in the United States of America

Published in 2023 in the UK by Profile Books

First US Edition: August 2023

Published by Basic Books, an imprint of Hachette Book Group, Inc. The Basic Books name and logo is a trademark of the Hachette Book Group.

Print book interior design by Six Red Marbles.

Library of Congress Cataloging-in-Publication Data
Names: Cheng, Eugenia, author.
Title: Is math real? : how simple questions lead us to mathematics' deepest truths / Eugenia Cheng.
Description: First US edition. | New York : Basic Books, [2023] | Published in 2023 in the UK by Profile Books. | Includes bibliographical references and index
Identifiers: LCCN 2022050481 | ISBN 9781541601826 (hardcover) | ISBN 9781541601840 (ebook)
Subjects: LCSH: Mathematics—Philosophy. | Mathematics—Popular works.
Classification: LCC QA8.4 .C436 2023 | DDC 510.1—dc23/eng20230509
LC record available at https://lccn.loc.gov/2022050481

ISBNs: 9781541601826 (hardcover), 9781541601840 (ebook)

LSC-C

Printing 2, 2023

To

everyone who has ever been made to feel bad at mathematics.

You didn't fail math: math failed you.

CONTENTS

INTRODUCTION

When I was in school, one of my favorite classes was the one in which we made stuffed animals. I made a fluffy poodle, and a sleeping puppy with soft, velvety ears. I loved the whole process, from cutting out the pieces, seeing how they miraculously fit together to make an animal, and sewing them together, to the magical moment of turning the whole thing inside out, and the joy of stuffing it so that it seemed to come alive.

Why make a stuffed animal when you could just buy one? Why make anything yourself when you can just get it ready-made instead?

Sometimes it's because the ones we make ourselves are better. I find homemade cakes much more delicious than bought ones. But sometimes the things we make ourselves aren't objectively "better." I enjoy playing the piano, although I can hear much "better" performances if I put on a recording or go to a concert. I even enjoy occasionally making my own clothes, although they're far from perfect.

Sometimes it's because it's cheaper. It's much cheaper for me to cut my own hair, so I do, even though a professional haircut would look "better."

But often, it's just satisfying to make something yourself. This is true for me for food, music, clothes, but different people find different things satisfying. Another variation on this theme is the idea of climbing a rock face just with your bare hands (no thanks), climbing Mount Everest without oxygen (also not for me), or rowing across the Atlantic (I'll pass on that too). Or perhaps it's like going on a camping expedition where you carry everything on your back including your food and your tent, so that you can spend a little while being self-sufficient out in the wild.

For me, math is also about making something myself: it's about making *truth* myself. It's about being self-sufficient out in the wild world of ideas. This, to me, is an immensely exciting, daunting, awe-inspiring, and ultimately joyful experience, and this is what I want to describe.

I want to describe what math *feels* like, in a way that is quite different from how it's often thought of. I will describe the expansive side of math, the creative, the imaginative, the exploratory, the part where we dream, follow our nose, listen to our gut instinct, and feel the joy of understanding, like sweeping away fog and seeing sunshine.

This is not a math textbook, nor is it a math history book. It's a math emotions book.

Math inspires rather different emotions in different people, and unfortunately for some people it mostly represents fear, and the memory of being made to feel stupid. I would like to show math in a different emotional light.

Some people love math and some people hate it, and unfortunately the way some of the math-lovers talk about it makes the other people hate it even more. The thing is that there are two very different reasons people love math. Some people like it because they think it has clear right-and-wrong answers. They find it easy to get the answers, and this makes them feel smart. Some people dislike it for more or less the same reason, but the other way around: it has clear right-and-wrong answers, but they find it hard to get the answers, and this makes them feel stupid. Or, most likely, they are made to feel stupid by people who get the answers more easily. And they don't even like the idea of clear answers. They see the subtle nuances of life and don't think that something so black-and-white can capture what they find most interesting about life.

However, this image of a rigid world with clear answers is a very limited view of what math is like. Abstract math really doesn't have such clear right-and-wrong answers, especially not at a research level, but only a small proportion of people ever make it to that stage to see what it's really like. And the extraordinary thing is that those mathematicians often love math for the same reasons that math-phobic people dislike it: they are interested in subtlety and nuance, to express and explore what is most interesting about life. Deep down, math isn't about clear answers, but about increasingly nuanced worlds in which we can explore different things being true.

So there is this curious effect: research mathematicians and math-phobic people have some similar attitudes toward math. It's just that for the former group, those attitudes are nurtured and celebrated, but for the latter, they are met with disdain or even ridicule, and the latter people may never find out how close their thoughts and feelings are to those of a research mathematician.

There's a gap between the reality of what math is, and the perception of what math is. I want to close that gap. Too many people are being put off math unnecessarily. Too many people are being made to feel stupid for asking questions about math that sound basic but are actually important and profound. They have burning questions, but they're told those are stupid questions, or that they're not supposed to ask those questions in math, when really the questions are mathematically very interesting. I want to answer those questions, but moreover, I want to validate and celebrate those feelings of wanting to understand more rather than take math for granted. This is important because the whole point of math is exactly that we're trying *not* to take things for granted.

My aim isn't to evangelize or to persuade everyone to love math. Different things motivate and put off different people, so there isn't one solution to getting people interested in math. My aim is just to shed some light on what math really is, to dispel myths about it, to clear up misunderstandings, and to stop putting people off for these particular wrong reasons. If you see the true nature of math and still don't like it, that's up to you—we don't all need to like the same things.

I just think it's a shame that so many people think they don't like math when they've only been shown some very narrow, unimaginative, authoritarian version of it, a version that doesn't allow for any personal input and curiosity of their own.

The feeling of personal input is very important to me. Sometimes, especially after a long teaching day, I just feel too tired to make myself dinner, even if most of my dinner is already made and I just need to open a package of pasta and cook it. However, somehow I am never too tired to make a cake. I've realized that this is all about personal input. I might have no energy to do something if I feel it involves no personal input and no creativity, while still having energy for something that looks like more effort, but involves personal input and creativity and so feels worthwhile to me.

And this is one reason people get put off math. If you like having personal and creative input, then routine math according to predesignated algorithms is not interesting and thus is too much effort. You might rather make a twelve-piece tiny thumbnail-sized tea service out of Play-Doh, as one art student did in my class when we had been using Play-Doh for a mathematical activity and I said they could continue to use it while we had a discussion.

When we teach math, there is a tension surrounding the different types of people we think are in the audience. There is a tension between wanting *some* students to be able to do everything "correctly" and accurately if they're going to go into research or jobs that depend on it, but also knowing that most people in standard math classes are not going to do that. Trying to get everyone to meet the standards needed for mathematical jobs would be like teaching children cooking as if they're all training to be line cooks in a professional kitchen. Instead it's better (in both cases, I suspect) to show them the possibilities, nurture enjoyment and curiosity, and trust that they can learn the more precise skills later if they need and want to.

Usually when I say things like this some people get up in arms and say, "But there are basic math skills that are crucial for everyday life!"

I suppose there are, but I don't think there are really that many, or they're not really extremely crucial. Most of the scenarios in which they are "crucial" are rather contrived. And either way, we are teaching plenty of things that aren't crucial at all, and we need to weigh that up against the harm we're doing by actively putting so many people off math with an unimaginative and limiting approach.

Math too often seems to come from rigidly imposed rules, which is why it inspires fear. But really, math comes from curiosity. It comes from instinctive human curiosity, and from people not being satisfied with answers and always wanting to understand more. It comes from questions.

Have you ever wanted to ask something about math but been told that's a stupid question? There are many innocent questions that even children ask, such as: Is math real? Where does it come from? How do we know it's right? Unfortunately too many people are discouraged from asking these questions. They are told that these questions are "stupid." But there aren't really any stupid questions in math. In fact, these "stupid" questions are the very same questions that mathematicians ask, that drive mathematical research, and push the boundaries of our mathematical understanding.

It might seem that math is about *answering* questions, but one of the most important parts of math is the *posing* of questions. In this book I'm going to show how those questions might sometimes seem innocent or vague, naive, simple or confused, but they can lead to some of the most profound mathematics that's out there. These questions align with qualities we often don't associate with math: creativity, imagination, rule-breaking, play.

We should encourage those types of questions, not suppress them.

If we give students the impression that they shouldn't ask those questions then we're giving them the impression that math is rigid and autocratic, and that we shouldn't question it. And that is the *opposite* of what math is. The whole point about math is that it's been built on rigorous foundations precisely in order to withstand deep questioning. All questioning. And when we can't answer questions, the mathematical impulse is not to suppress those questions, but to do more math in order to be able to answer them.

This is why questions lead to deep math.

When you start research as a PhD student, one of the hardest things is figuring out what is a good question to work on, and this is often one of the most important roles that a supervisor plays. In my research in the abstract field of category theory, often most of the work is in working out exactly what question we are asking in the first place. In school math, we put too much emphasis on *answering* questions rather than on asking them. I tried to look up "great questions children ask in math" online, and alas all I got were great questions to *ask* children about math. It's as if all the resources out there think *we* should be asking the questions and children should be answering them. This is the wrong way around.

I want to encourage and validate the asking of questions, the ones you've always wanted to ask but which were never answered, the ones people said were not the point, the ones that made people say you should just buckle down and do your homework. The ones that made you feel you weren't a "math person," because the people who did well in tests didn't seem to be asking *those* questions. The ones that slowed you down because you weren't satisfied with just writing down the answers you were supposed to write down. This book is about those questions, because those questions are fundamental to the deepest forms of mathematics. That's not the kind of math where we add and multiply numbers to get an answer, or where we calculate angles in a triangle, or areas of some random shape, or solve some meaningless equation for no other reason except to get a good grade on a test. I mean the kind of math that drives researchers at the forefront of abstract math. The kind of math that takes half a lifetime for someone to understand, or maybe hundreds or thousands of years, and even then we still only partially understand it. The kind of math that doesn't seem to have any direct application to our daily lives, that doesn't immediately solve some problem in life or enable some new machine to be built. The kind of math that exists mainly in our heads.

Is this kind of math real?

Were the fluffy puppies I made at school real? Well, they weren't real puppies, but they were real stuffed animals.

Math isn't exactly "real life," but it's still real. It's real ideas, it's real thoughts, and it produces real understanding. I love the clarity it gives me, and regret that this can sometimes seem like it's turning everything into rigid black and white, rather than shedding light on ambiguity. But I sympathize with anyone who has that impression, because of how math too often comes across. I experienced that in school math too.

Here is a graph of my love of mathematics over time, or rather, my love of math *classes* over time:

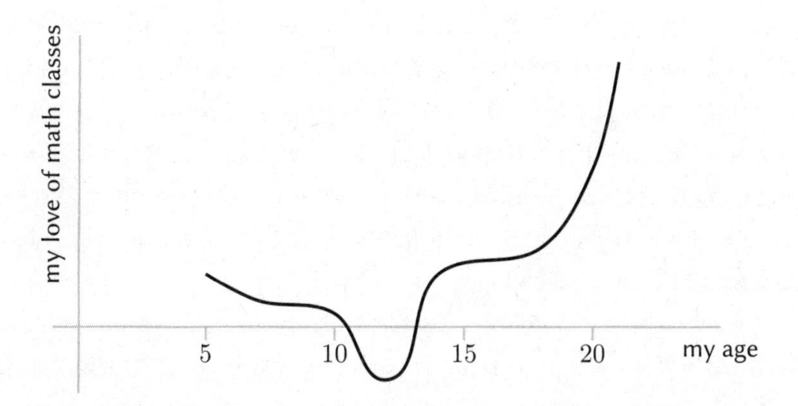

I quite liked math, as far as I knew what it was, when I was five. It was a bit of continuous deterioration through elementary school though, and then I reached a point in middle school when I actively disliked math lessons as I found them tedious and pedantic. I don't blame my teachers at all; I blame the curriculum and the exam system. The situation improved when I started doing slightly more advanced math, especially the investigation component of high school math in the UK, which was the only part I liked. These were open-ended projects we did across several weeks of the term, which started with a fairly structured question and then opened out into limitless possibilities for independent exploration. Then finally when I was doing Further Math A-level[†] I really enjoyed some parts of pure math: especially abstract algebra, proof by induction, and polar coordinates (I'll talk about those later in this book). It wasn't until I got to college that things really picked up, and when I started my PhD my love of math went off the charts again, although that's also when we stopped having classes and learned things by reading, discussing, and going to seminars.

But actually my love of *math* across time was constant. I could depict it in the graph as a constant horizontal line, high up above the wiggly line showing my feeling about math classes. I was lucky that my mother showed me fun and exciting and mysterious and mind-boggling things about math completely separate from anything we had to do at school. That instilled in me a belief that math was more than what was in the curriculum, and that it was something fun and exciting and mysterious and mind-boggling. My love for that never wavered. I know that most people do not have a mother at home showing them those things and answering their innocent questions, and so their love of math might plummet as mine did, but never recover. This is what I hope we can change.

I want to help math-phobic people overcome their fear (and perhaps trauma) and see why research mathematicians love math, which is very different from just "loving numbers" or enjoying getting the right answer. I want to show that the reasons people are put off math are really unfortunate and not based in the true nature of math. I want to demonstrate that the innocent, open-ended, "stupid-sounding" questions about math are valid, that they are good questions, and that questions like that are

† A-levels are the public exams in the UK at the end of high school. In my day students chose three or four subjects, but math could be done in a double quantity, with Further Math being the more advanced part.

essential to mathematics. I want to persuade you that if you think you're bad at math or were labeled as bad at math at school, it's possible you were just seeking a deeper level of understanding and that nobody helped you reach it. And I want to give an idea of what it's like to do research math, to explore the world of math and uncover deeper and deeper truths as we get further and further into the mysterious undergrowth.

I will start each chapter with one of those innocent questions that might sometimes be deemed a "stupid question," but I will show how the process of digging deeply into the question leads us to important mathematics and often entire fields of research. It's a slow process of gradually probing and pulling back the undergrowth to see what's behind it, and it can sometimes seem like we have to take several steps backwards in order to see what we're doing. It can also seem like we're taking very small steps and not getting anywhere, before suddenly we look behind us and discover we've climbed a giant mountain. All these things can be disconcerting, but accepting a little intellectual discomfort (or sometimes a lot of it) is an important part of making progress in math. The discomfort is often a starting point for development and growth. Sometimes it feels like a kind of vertigo as we notice a chasm between what our intuition tells us and what our careful logic tells us, or perhaps a chasm between two different gut feelings, like when I saw friends for the first time after two years of pandemic isolation, and it simultaneously felt like no time at all and also several lifetimes.

I will start with broad ideas about math, and then gradually zoom in, first to particular subjects in math, and then to individual tales. So the first four chapters will be about the general idea of math: where it comes from (Chapter 1), how it works (Chapter 2), why we do it (Chapter 3), and what makes it good (Chapter 4). Then I'll move onto talking about specific aspects of math: the use of letters (Chapter 5), formulas (Chapter 6), and pictures (Chapter 7).

I'll end in Chapter 8 with some individual stories starting from naive questions, and show how a mathematician relates a question to existing knowledge, often finding answers from deep parts of research math.

I will try to depict what it feels like to be following your nose through meandering mathematical thoughts, rather than being marched through them on command and with a time limit. It's the difference between being taken to the other side of a forest because you need to get there, and learning how to cross a forest so that you know how to do it and, moreover, can appreciate the creatures and the undergrowth that you see along the way. The former has its place, but the latter has a broader purpose. It's also longer and harder, but has the potential to be much more satisfying. I certainly find exploring that landscape myself more satisfying, and indeed enjoyable, as well as much more illuminating. And, as I'll explain, it's even useful to me in my daily life, useful in a much broader sense than for specific tasks like splitting a bill or doing my taxes: it helps me to think more clearly about anything and everything.

Perhaps there is a tension between clearly defined, specific uses, which are easy to describe, and broader, more general applicability, which is harder to pin down. But just because it's harder to pin down doesn't mean we should dismiss it; on the contrary, perhaps the things that are hard to pin down are the most worthwhile. These are not the easily memorized, easily recited facts about mathematics, but the deep truths.

So this book is about the deep truths of mathematics, but more to the point, it's about how we arrive at the deep truths of mathematics. The deep truths themselves are important, but really I want to show how we get to them. It's much like that old saying about teaching someone how to fish rather than giving them a fish: if I tell you some deep truths of math, then all I've given you is those truths. But if I show you how to access mathematical truth, then all of mathematical truth becomes available to you. At one level, this book is about some specific questions and answers, but at a deeper, more important level, it's about how questions lead us on a journey, about where the journey is leading, and about why we might want to go on that journey, and what we see on the way.

Math might seem like it's about getting the right answers, but really it's about the process of discovering, the process of exploration, the journey toward mathematical truth, and how to recognize when we've found it. That journey starts with curiosity, and curiosity makes itself known in the form of questions.

CHAPTER 1

WHERE MATH COMES FROM

Why does 1 + 1 = 2?

One possible answer to this question is "It just does!" That is really a variation on "Because I say so!" an answer that has been frustrating children for generations. "Because I say so" means that there is an authority figure who makes the rules, that they don't have to justify their rules but can make up any rules they want, and everyone else is a minion who just has to follow those rules.

It is quite right to feel frustrated by that idea. In fact, a strong mathematical impulse is to immediately want to break all the rules, or find sneaky situations in which those rules don't hold, to show that the supposed authority figure doesn't have quite as much authority as they think.

Math can seem like a world of rules you just have to follow, which makes it seem rigid and boring. By contrast, my love of math is somewhat driven by my love of breaking rules, or at least pushing against them. I'm a bit sheepish about that as it makes me sound like an adolescent who never grew up. My love of math is also driven by my wanting to keep asking "Why?" about everything, which in turn makes me feel like a toddler who never grew up. But both of those impulses play an important role in advancing human understanding, and in particular mathematical understanding. Those impulses are an important part of the origins of math, which is what we'll be looking at in this chapter.

I'd like to stress that in normal life I am a very law-abiding person because I understand rules that are about holding a community together and keeping people safe. I believe in those rules. I don't mind following rules that have a purpose; the rules I don't believe in are the arbitrary ones that don't seem to have a justification, or whose justification I don't believe in. Rules like "You must make your bed every day" (which really isn't to my taste) or "Never melt chocolate in a microwave" (it's certainly easy to ruin it, but as long as you stir it every fifteen seconds without fail I've found it's fine).

So I want to look at where the apparent "rules" of math come from, and indeed where math comes from at all. I'll describe how it starts from small seeds and then grows to great heights in an organic way. The seeds are naive questions that any of us might pose, and that small children often pose innocently, like when they wonder why 1 + 1 is 2, rather than just being content to know that it is. Like any seeds, they need to be nurtured in the right way to grow. They need fertile soil, space to plant their roots, and then nourishment. Unfortunately our innocent questions are too often not nurtured in this way, but dismissed as "stupid" and tossed aside. But the difference between deep mathematical questions and innocent ones might only be the nurturing—that is, there is no difference. It's the same seeds.

People who don't like math are often put off by the apparently autocratic declaration that something is the right answer, without explanation. "One plus one just *is* two." But wondering why something is true leads us to build strong foundations for mathematics, so that we can make clear and rigorous arguments. Some people find this clarity and reliability relaxing and liberating, while others find it restrictive and autocratic. A question like "Why does 1 + 1 = 2?" allows us to explore the idea that math doesn't have clear right answers, but rather, different contexts in which different things can be true. This is going to lead us to explore where numbers come from in the first place, how we come to the ideas of arithmetic, and how we can then use those ideas in other mathematical contexts such as when we're thinking about shapes. This touches on many important themes in how math is developed, starting with making connections between things, taking abstraction seriously, and then expanding our thought processes to encompass more of the world around us, little by little.

So rather than think about why one plus one is two, let's go a little further and question whether it's even true all the time.

Pushing against boundaries

Children seem to be natural seekers of counterexamples. A counterexample is an example showing that something isn't true. Declaring that something is always true is like putting a boundary around something, and seeking examples that contradict that is like pushing against those imposed boundaries. It's an important mathematical urge.

You can try prodding a child on one plus one by saying something like this: "If I give you one cupcake and another cupcake, how many cupcakes will you have?" But they might gleefully declare "None, because I've eaten them!" or indeed, "None, because I don't like cupcakes." I am always delighted when I see parents posting pictures of their children's defiant answers online. One all-time favorite of mine was in answer to the question "Joe has 7 apples and uses 5 of them to make an apple pie. How many apples does he have left?" to which my friend's child wrote "HAS HE EATEN THE PIE YET?" I enjoy answers that are arguably correct while definitely not being the answer that is supposed to count as correct. This shows an important aspect of mathematics, and the children's thought processes are showing an important but underappreciated aspect of mathematical instinct, the instinct to push against unjustified authority.

Children may be trying to prod at authority because they're exploring the boundaries of situations, or because they're trying to find a sense of self in a world that gives them very little control over anything at all. I remember being a child very clearly, and how frustrating it was that I had to do what adults told me all the time. If I could sense that an adult was asking me a leading question it was rather fun to veer off in a different direction, like by saying I didn't like cupcakes.

That is, in a way, a cheeky and disruptive urge, but I also think it's a mathematical urge. Yes, perhaps mathematics is cheeky and disruptive, but another way of putting it is that mathematics is seeking the boundaries of things, just like children are. We want to be clear about the limits to when things are true, so that we can be sure when we're in a "safe" area, but also explore outside that area if we're feeling daring or curious. It's like a toddler running off into the distance to see at what point an adult will come chasing after them. Thinking about situations in which one plus one is not two is an example of that.

I have written before about situations in which one plus one is actually zero. If I say "I'm not not tired," that means I'm tired, and some children find it very funny to say "I'm not not not not not not not not not not not tired!" and dissolve into hysterics because they know nobody has been able to keep track of how many times they said "not." The point here is that one "not" plus one "not" is the same as zero "nots." This reminds me of some terrible exam question I once had to grade that involved a long and tedious calculation with many moments for potentially getting a negative sign wrong. It was particularly painful to grade because if students made two such errors then the answer would turn out right, or indeed if they made four. However, in math the answer doesn't really count as right unless the process is right (a point I'll come back to in the next chapter) and so I had to look rather hard to see if the process was wrong even though the answer was "right."

Another situation in which one plus one is zero is when everything is already zero, like the world of candy I lived in when I was little: I was allergic to artificial food coloring, and all candy had artificial food coloring in those days, so no matter how many candies I had I effectively had zero.

Sometimes one plus one can equal more than two because of rounding errors. If we're only working with whole numbers then 1.4 counts as 1 (to the nearest whole number); but then if you do that twice you'll get 2.8, which counts as 3 (to the nearest whole number). So in the rounded world it will look like one plus one is three. A related but slightly different situation is if you have enough cash on you to buy one cup of coffee, and your friend has enough to buy one, then together you still might have enough to buy three, because even if you have one and a half or even 1.9 times the money needed for a cup of coffee, that still only gets you one cup of coffee on your own.

Sometimes one plus one can equal more than two because of reproduction: say perhaps you put one rabbit and another rabbit together, then you might well end up with rather a lot of rabbits. Or sometimes it's because the things you're adding together are more complicated: if one pair of tennis players gets together with another pair of tennis players for an afternoon of tennis, there ends up being more than two pairs of tennis players because they could play each other in all sorts of different combinations. If the first pair are called A and B, and the second pair are called C and D, then we have the following pairs in total: AB, AC, AD, BC, BD, CD. So one pair of tennis players plus another pair makes six.

Sometimes one plus one is just one, like if you put a pile of sand on top of another pile of sand then you just get one pile of sand. Or, as an art student of mine pointed out, if you mix one color with one color you get one color. Or, as I saw in an amusing meme, if you put a lasagna on top of another lasagna it's still just one lasagna (a taller one).

A slightly different scenario in which one plus one is one is if you have a voucher for, say, a donut with your coffee, but there's a maximum of one per person, so that even if you have another one it's just like you only have one (unless there's someone you can give it to). Or if you're pressing the OPEN DOORS button on a train, pressing it more than once is just the same as pressing it once. At least, it's the same in terms of the effect it has on the doors, but perhaps not in terms of the amount of frustration you get to express; perhaps that's why people so often seem to stand there hitting it repeatedly.

Now, you might think that the situations above aren't *really* situations in which one plus one equals something else, because they aren't *really* addition, or those aren't *really* numbers, or some other reason that they don't count. You're welcome to think that, but that's not what math does.

Math instead says: Let's work out what those contexts are and what they mean. Let's work out what the consequences are of things being like that, and see what other contexts we can find in which things work in similar ways. Let's be more clear about the context in which one plus one really does equal two, and contexts in which it doesn't, and in doing so we'll understand something about the world more deeply than we did before.

This is where math comes from. And in order to explore contexts in which one plus one does and doesn't equal two, I want to do more than just dig into where that equation comes from. I want to dig all the way into where any math comes from at all.

The origins of math

Math comes from wanting to understand things better. And in order to understand things better we find a way to think about them that makes them easier. One way to make things easier is just to ignore the hard parts, but a better way is to come up with a point of view that enables us to focus on the part that is relevant to us right now, while not completely forgetting that the other parts exist.

It's a bit like having a filter over a camera lens that temporarily enables us to focus on particular colors, before switching it for another filter to look at other colors. Or like that stage of making a stew where you strain the liquid out in order to reduce and thicken it—you don't throw away the stuff that you've strained out, you put it back in again afterward.

The most well-known basic starting point for math is numbers. This is most children's first introduction to math and their first impression of what math is, and it remains many people's lasting impression of what math is. But math is about so much more than numbers. And even when math appears to be about numbers, it's often not *actually* about numbers, but rather, about the process of how we get from our world to the world of numbers at all, and what insights we gain by doing so.

The thing about the strong association of math with numbers is that numbers can seem boring to anyone who likes ambiguity, creativity, wild exploration, imagination. I'm not going to try and claim that numbers are interesting, but the opposite: that they are indeed boring, and that that's the entire point of them.

The point is to encapsulate one aspect of the world around us so that we can be done with that part as fast as possible, leaving the more interesting part of our brain to deal with the more exciting parts of the world. It's like getting a computer to run all the least interesting parts of life (which for me is paying bills, ordering groceries, scaling recipes) so that I can save my brain for the more interesting parts: interacting with people, playing music, cooking delicious food.

Numbers come from wanting to simplify the world around us. No wonder the result is simple, and thus, potentially boring. But how we invent numbers in the first place is rather profound. Numbers come from us making an analogy between different situations, and choosing which part of those situations to ignore temporarily. We might look at two apples and two bananas and see some similarity between them, which we then encapsulate in our brains as the concept of "two." But in order to do that we have to ignore the applehood of the apples and the bananahood of the bananas, and see those things as just objects, abstractly, without those particular properties.

This is a leap of abstraction. It's a difficult leap to make, and no wonder it takes children a while to make that leap. We can encourage them to do it, by repeatedly counting things right in front of them, but in the end they have to make that leap for themselves, and we can't do it for them.

The trouble is that it can seem reductive to forget those crucial features of the objects in question, and if we only focus on the part where we "made things more boring" then we make everything sound boring, instead of focusing on what the point of it was, which was to open up amazing new channels of understanding.

The point of abstraction

By inventing numbers we've done something quite profound: we've made an abstraction. Abstraction is when we forget some details about a situation in order to consider an "idealized" version, which is not the same as the concrete (real-world) version, but captures the features we want to think about right now. It takes us further away from the real-world situation, but with an aim: the aim is to find similarities between different situations so that we can understand more situations at once without having to do so much work. In a way, we're simplifying our building blocks in order to be able to build more creatively. It's like the difference between a jigsaw puzzle that has to fit together in a certain way to make a given picture (and the pieces will only fit together one way) and a jigsaw puzzle with pieces that are more general, that fit together in many different ways, where the aim is not to see if you can build one particular picture, but to see all the different structures you can build. I always enjoyed the tangram puzzle for this reason, the pieces just being generic geometrical shapes: a square, some triangles in different sizes, and a parallelogram. It is thought to date back to eighteenth-century China, although similar constructions date back to much earlier Chinese mathematicians. The shapes can fit together into a square as shown below, but can also be used to make an endless range of other shapes, depicting people, animals, or whatever you can imagine, albeit in a somewhat stylized way. Here's a rabbit:

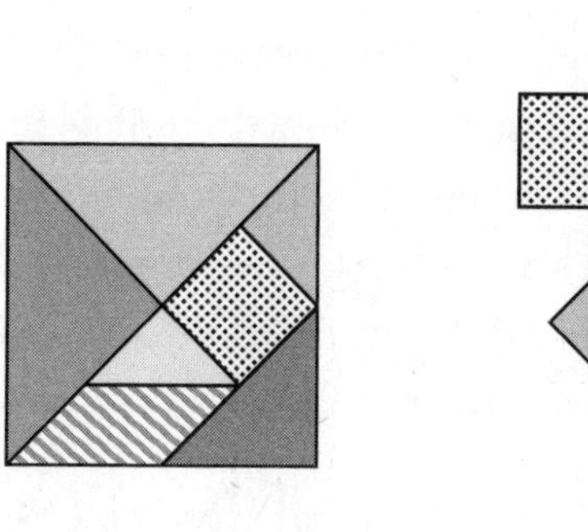

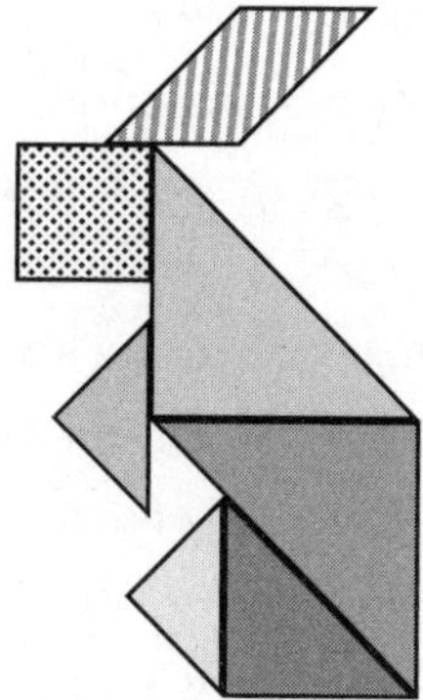

Numbers are also a way of opening up such a world of endless possibilities, just in a rather less visually vivid way. (We'll come back to the use of pictures in math in Chapter 7.) Aside from being less pictorially enticing, numbers can seem rather closed off if the only thing you then do with them is answer specific questions with specific answers, and then you get told you're either right or wrong.

Numbers are definitely not the be-all and end-all of mathematics, but they are a beginning of how we learn to reason with abstraction. The important steps of this process are something like the following.

First we decide what aspects of a situation we're interested in. Perhaps we spot similarities between various different situations and feel curious about why those similarities are arising. Then we perform the abstraction: we focus on the parts of the situation that are similar, and if we're focusing on quantity, we come up with something like numbers, which is the "essence" of what we are focusing on right now. This constitutes the creation of a new abstract world, which we can then explore to work out how things work there, what sorts of creatures live there, what sorts of weird and wonderful landscapes lurk around the corner.

If we ever feel constrained by that world, we can just make a new one and explore that, and we often do. On the other hand, if we feel like we want to know more about how this relates back to the world around us, we can also do that. If we want to make a different relation between the world around us and our abstract one, we can also do that. For example, we can measure quantity in different ways, count things differently, or associate numbers to things in many different ways like when we rate a restaurant according to various different criteria. If we want to focus on a different aspect of the world around us, such as shape instead of quantity, we can do that.

It's a bit like being given a new box of paints, and deciding to mix them up a bit to see what you can do with them. But the wonderful thing about mathematical paint is that you never run out of it. There's no risk that you'll "waste" it by mixing it in ways you don't like, because they're just ideas, so there are always more around to experiment with. Playing around with numbers doesn't use them up, and this is the same of all abstract concepts. It is one of the most fun and satisfying aspects of math, to me. But it does also raise the perplexing question of whether any of this is real.

Are abstract concepts real?

The first thing I think about this question is: What does "real" even mean? Is anything real? If I think about it too hard I can quite easily convince myself that I'm not real and that nothing is real.

If you've ever wondered if math is real, you might have been told it's a stupid question. Perhaps you looked around you and saw that people who were "good at math" didn't worry about this sort of question, they just got on with getting the right answers.

Well, I'd like to reassure you that mathematicians and especially philosophers do wonder about the status of math. Do numbers exist? I am not a philosopher so I am not going to go into the philosophy of that, I am just going to say what I think.

In order to examine what it means for something to be "real" it might help us to think of some things that we think of as real and not real. There are many concrete things we can touch that we probably all agree are real. The world is real, people are real, food is real. Then there are things we can't touch that we might think are real, such as hunger, love, poverty. Then there are perhaps things we can agree are not real, such as the Easter Bunny, the Tooth Fairy, and Santa. And there are some things that people really don't agree on, like God, UFOs, ghosts, and unfortunately COVID.

Hold on a second though, because actually I believe that Santa and the Tooth Fairy *are* real. At this point you might think I've lost my mind, but let me try and explain.

Small children in certain cultures think (or are told by adults) that Santa or Father Christmas is a man with a fluffy white beard who wears a red outfit and flies around the world delivering presents to children, in a sleigh drawn by reindeer. At some point they grow up and are disillusioned to discover that actually their presents (if they celebrate Christmas) come from their parents, and are simply placed under the tree while the children are asleep. This is thought of as children realizing that Santa "isn't real."

However, I argue that this doesn't mean Santa isn't real, it just means that the rather unrealistic traditional description of Santa isn't literally accurate. However, *something* exists: there is something that results in presents being delivered to children all around the world on Christmas morning. That thing is an abstract concept—it's an idea of Santa. You might think that the idea of Santa exists, but Santa still doesn't. However, mathematical concepts are so abstract that they are nothing but ideas. The idea of the number two *is* the number two. And those ideas are real. I'm very used to treating abstract mathematical ideas as real objects, so I'm happy to think of Santa as a real abstract concept too. Taking ideas seriously and treating them as real things is an important part of how math develops.

How math develops

Math might seem like it's all about numbers and equations, but even if you think back to early school math you might remember that other things were involved, perhaps shapes and patterns, and pictorial representations like bar charts and Venn diagrams. The research I do (in the very abstract branch of math called category theory) doesn't involve numbers and equations at all. So if math isn't just the study of numbers and equations, what is it? I often like to characterize math as "the study of how things work," except it's not the study of any old things, and it's not any old study either. I say:

> Mathematics is the logical study of how logical things work.

The first issue here is that nothing really *is* logical: everything in life works by a mixture of logic and other things, such as randomness, chaos, emotions. Or possibly another point of view is that those things are also logical, just too complex for us to understand using logic. For example, the weather is actually logical, it's just that we will never be able to take accurate enough measurements of what's going on in the atmosphere to be able to use logic to predict the weather with any great certainty. The weather isn't illogical, it's just hard.

Mostly math deals with this by the process I was just describing: abstraction. We forget certain details about situations with the aim of moving us away from the messy "real" world into the abstract world of ideas, where things do behave according to logic, because we've conveniently ignored (temporarily) the parts that don't. I don't want to call the non-abstract world "real" though, because I don't think abstract ideas are unreal, so I prefer to call the non-abstract world "concrete": the world we can touch.

One fascinating aspect of math is that it isn't just defined by what it studies. Most subjects, like history, biology, psychology, economics, are defined by what they study, and then techniques are developed to study those things. But math has a cyclical situation going on, where *what* we can study is defined by *how* we study it, so that we can find new things to study as well as us then finding new ways to study those things. It's something like this:

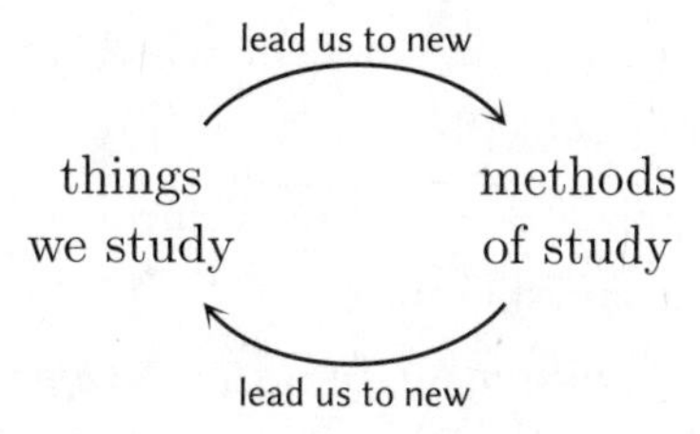

But really each arrow is giving us new things, so we're not going around the same circle over and over again—it's more like a spiral, where we are going around and around but also going up and up at the same time. We keep finding things to study using our new methods, and finding new methods for studying those things, and the situation keeps escalating, a bit like climbing up this spiral "staircase," starting at the bottom with numbers:

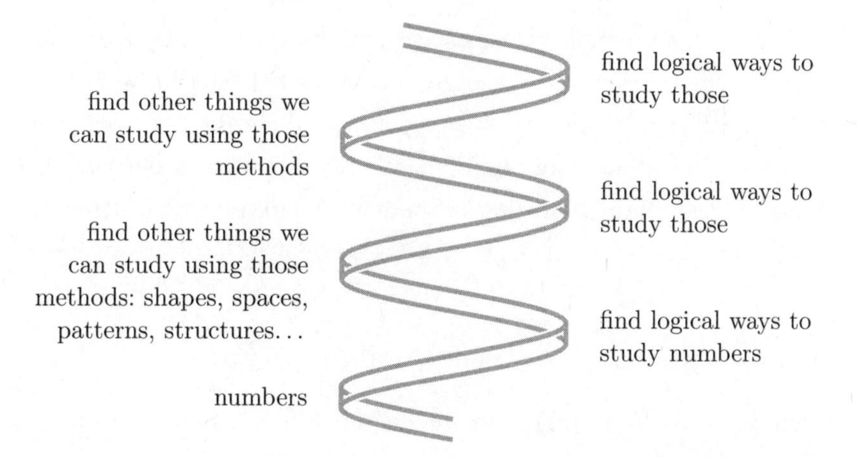

So we have a sort of spiral staircase going up forever. I'd like to demonstrate how this might work, by going on a little wander up the staircase to see where it can take us. This isn't directly addressing the question about one plus one, but it's doing some background exploration so that we can eventually address the question in a meaningful way.

The "staircase" all started by us working out how to perform abstraction on things around us to get to numbers. Numbers behave more logically than cookies, cows, or whatever we were trying to count. We then come up with some ways to study numbers, such as investigating how they are related to each other by adding and subtracting, multiplying and dividing.

We then realize that there might be more things we can study in a similar way, like perhaps shapes. We realize that we can perform abstraction on things around us to find similarities between, say, a window, a door and a tabletop, because they're all rectangles. And we find we can roll up a semicircle to make a cone for a hat, which is also a good shape for a traffic cone, or (if we turn it the other way up) a handy edible container for ice cream. Incidentally, here's what I mean by this being an emotional account of math rather than a history of math—of course, cones were studied long before traffic cones and ice cream cones.

So how do we study shapes using the techniques we developed for studying numbers? Well, we might think about adding and subtracting shapes, that is, sticking them together and cutting parts out. We can also think about multiplying shapes, but this is a little harder, and it pushes us to think more deeply about what multiplication *means*, as we'll now see.

Expanding the concept of multiplication

Multiplication of numbers might be something we're comfortable with or horrified by, but either way it's something we are likely to take for granted. However, if instead of taking something for granted we think hard about what it really means or how it really works, that sometimes enables us to get much more out of it. This isn't always the case—sometimes it just means we get mentally paralyzed, like if we think too hard about what life means. But if I think really hard about, say, my comfort zone, then I can get much more from it and can expand what I'm comfortable doing, rather than having to feel like I'm stepping outside my comfort zone.

We're going to think hard about multiplication of numbers, and this is going to lead us to a way of multiplying other things, such as shapes, thereby expanding the concept of multiplication. Mathematicians have pushed this further and further and have gradually come up with a whole theory of when it is possible to multiply anything at all. Abstract mathematicians do this for all sorts of mathematical concepts, not just arithmetic.

When we start out multiplying numbers, we might think of 4 × 2 as "four sets of two," that is, 2 + 2 + 2 + 2. We can take two counters, and then two more, and two more, and two more, and then just count them up to see how many there are in all.

This might make sense for multiplying numbers by shapes, because "4 × a circle" could just be four circles, but it doesn't make sense for multiplying shapes by shapes. If we were going to multiply a square by a circle, what on earth would "circle sets of square" mean? Well, we can interpret it to mean something if we're a little more expansive or imaginative about our multiplication. One way is to rethink 4 × 2, this time starting with a column of 2 counters, and "replicating" them along a line of 4 counters, making a 4-by-2 grid like this:

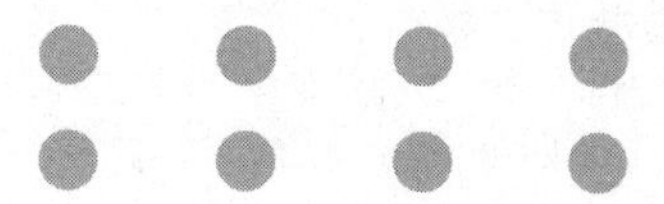

With some imagination we could now picture doing this to a circle and square—we wave a square around along a circular path, and keep replicating the square as we go. It's a bit hard to draw this on a page, but perhaps we could pick some different shapes, say a line and a circle. If we start with a vertical circle and "replicate" it through the air in a straight line, we make this shape: a cylinder.

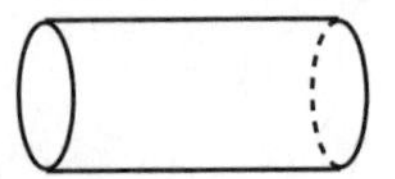

If we do it the other way around we're waving a straight line through the air in the shape of a circle, and we also make a cylinder. (In both cases we need to make sure we're waving in a direction that's sort of at right angles to our shape.) This is like two different schemes for knitting the sleeve of a sweater: we could use circular needles and build up circles on circles until we have a cylindrical sleeve, or we could use normal needles, knit a rectangle, and then join the seam up to make a sleeve. (I know, it would need to be wider at one end than the other to make a shapely sleeve, but I hope you get the picture.)

This is very vague, but gives the general idea of how we think of "multiplying shapes" together. What I've just said is that

$$\text{circle} \times \text{line} = \text{cylinder}$$

and also

$$\text{line} \times \text{circle} = \text{cylinder}$$

so we have something like

$$\text{circle} \times \text{line} = \text{line} \times \text{circle}$$

which is analogous to the more familiar

$$4 \times 2 = 2 \times 4$$

otherwise known as the "commutativity of multiplication." It's an example of how we can start studying shapes a bit like how we study numbers. We'll come back to more ways that happens later, including the idea of trying to understand what the fundamental building blocks are in each world.

Now that we've had the idea of including shapes in our logical studies, we can think of more ways of studying shapes, taking us another level up our "spiral staircase."

Continuing up the spiral staircase

We've looked at a way of studying shapes inspired by multiplication of numbers, but another way that isn't so relevant to numbers is symmetry. Shapes are more subtle than numbers, and symmetry is an aspect of that subtlety.

For example, a square and a rectangle are similar in some ways, and different in some other ways, and one way we can pin down the difference is to think about symmetry.

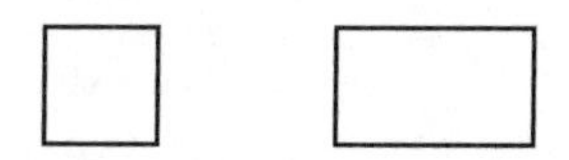

A square has more symmetry than a rectangle, because we can fold a square along a diagonal and the two sides meet up, but this doesn't work for a general rectangle.† In fact, this gives a handy way of making a square out of a rectangular piece of paper without having to use a ruler: fold a corner down as in the following diagram. We're essentially using the symmetry of a square (in principle) to make an actual square.

† I say "general rectangle" to be sure we're talking about a rectangle that is not in fact a square. In math a square counts as a special kind of rectangle, but in normal language if we were holding a square piece of paper and we called it a rectangle it would be a bit odd, like referring to someone as a high school graduate when they have a PhD.

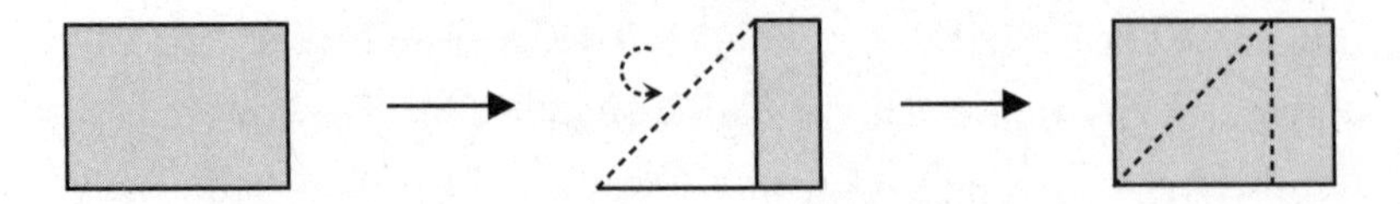

We can move on and develop our understanding of symmetry by including other types. The "folding" type is called *reflectional* symmetry because it's a bit like reflecting one part of the shape onto another as if with a mirror. Another type of symmetry is more like a windmill, where you can rotate a shape so that one part lands on another. This is called rotational symmetry.

Squares and rectangles have both types of symmetry, and we can then even think of combining symmetries to see what happens. This leads to the topic of *group theory*, which makes abstract structures out of symmetries and how they are combined. Once we are there, we realize there are other things we can study in a similar way, that are a bit like symmetry, but not involving shapes, taking us another level up our spiral staircase. One example is symmetry in words, that is, palindromes such as:

Madam, I'm Adam
A man, a plan, a canal, Panama.
Taco cat

We're quite good at recognizing that these can be read backwards, producing the same thing as reading them forwards, even though it involves a little bit of ignoring of spaces and punctuation. There's a type of symmetry in equations too. For example, if you look at this expression,

$$a^2 + ab + b^2$$

the a and the b are playing analogous roles in there. That is, if we swap a and b everywhere in that expression we get

$$b^2 + ba + a^2$$

which is the same as the first expression (provided the order in which we do addition and multiplication doesn't matter). This is another type of symmetry that is studied in a field called *Galois theory*. So we've moved on to including expressions involving letters in our math. (If the letters make you nervous, I sympathize; we'll come back to what the letters are doing there later.)

We then develop more ways to think about those expressions involving letters, maybe by thinking really hard about the relationships they might have with shapes. So where previously we studied shapes, and we also studied expressions in letters, we now think about the relationship between those things. This leads us to my research field, the field of category theory. This is a subject that focuses on relationships between things, and pushes that idea further and further so that we can study relationships between almost anything. Moreover, we can start to *regard* things as "relationships," even if they weren't originally relationships, in order to study them in a similar way. For example, we can regard symmetry as a relationship between an object and itself, which might sound bizarre but it turns out to be a very fruitful little bit of mental gymnastics.

And this is a crucial point: that because math starts with abstraction, we can study more things mathematically if we just think of new ways to perform an abstraction. This will bring more examples into our analogy that previously didn't seem to be there. This is different from studying, say, dolphins, where you can't exactly go around regarding things as dolphins in order to study them as dolphins if they aren't already dolphins. But with an abstract concept like a relationship, you can do that. We can regard symmetry as a relationship between an object and itself. We can regard train journeys as a relationship between the starting point and the destination. We can regard numbers as relationships between other numbers, such as 3 being a relationship between 2 and 5 since it's the difference between them.

In this way, the starting point for math is the mental gymnastics of finding ways to think flexibly about situations, in order to make connections between things that didn't previously seem related. And having a vivid and creative imagination is very helpful in conjuring those connections into being.

Making connections

Abstraction sounds like a process of moving further away from the concrete world, but really it's a way of making analogies between things, that is, finding connections. I really love finding connections between things. I love making connections between people. I love it when pieces of music remind me of other pieces of music. I love it when it dawns on me that an actor in one film is the same as one in another film I've seen, when it's not very obvious, like Crispin Bonham-Carter who played Mr. Bingley in the BBC *Pride and Prejudice*, and then popped up in *Casino Royale*. I particularly love spotting similarities between situations and realizing it means I've already understood this situation in a different context, so I don't have to start from scratch. This is how Miss Marple solves murder mysteries in Agatha Christie's books, and I enjoy those too.

In life we are often quite wrapped up with finding differences between things. We emphasize differences between people's experiences to make sure that we are not treating everyone of the same race as a monolith, or assuming that all women vote the same way. We point out not just the ways in which someone is in an oppressed minority, but also the ways in which different oppressed minorities are oppressed differently, especially those in the intersection of several minorities, to make sure we are not erasing people's experiences.

That is all important, but it's also important not to lose sight of connections between us. In fact, I'm convinced that strengthening our connections is crucial if we want to loosen the grip that white patriarchy has over society. If minorities fragment into more and more disparate groups this works in favor of the white patriarchy, who can only hold onto power if individual minorities do not work together to change those power structures. If individual minorities build enough connections to work together, then they become a majority.

In math we don't do one or the other of these things, but we remain ever flexible in our thinking: we point out senses in which things are connected, and also senses in which things are different. And we don't remain fixed in that point of view, we just take that point of view to see what we can learn from it, and then take another point of view to see what we can learn from that. This is a very different feeling of "doing math" from the one that feels like a rigid system of rules you have to follow.

Finding a sense in which things are the same is a starting point, a launchpad for us to be able to study different things at once, as when we first move into the world of numbers.

Another example is with shapes, where we think about shapes that aren't exactly the same, but perhaps are just scaled versions of each other, like these two triangles:

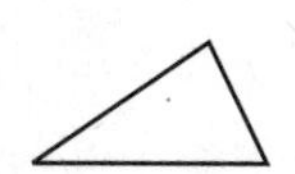 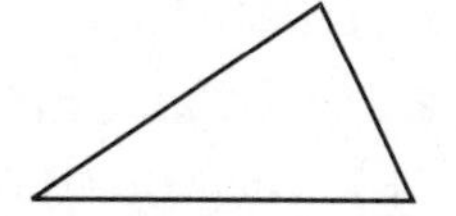

In some situations, we need to count triangles as the same only if they're exactly the same in every way. For example, if we're fitting pieces together, then only the exact correct triangle will fit. This gives us the version of sameness called "congruence."

In other situations, it doesn't matter how big the triangle is, for example, if we're just trying to calculate an angle, or if we're scaling an entire situation. So we get the notion of *similar* triangles, which can be scaled versions of each other like the ones above—it just matters that they are scaled entirely in proportion, which results in them having the same angles, and the same proportions between their three sides.

In yet other situations, it doesn't even matter what shape the triangle is, as long as it's some kind of triangle. For example if we're just trying to make a rectangular picture frame rigid, we need to put some crossbars on the back to make a triangle shape at each corner. It doesn't matter what shape the triangles are.

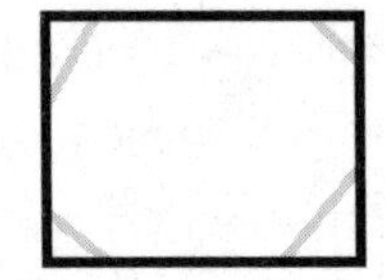

This is how we get to the notion of "triangle" in the first place: it's just a shape made from three straight edges, thus having three angles.

The study of congruent and similar triangles might seem pointless and contrived if we're just asking whether or not triangles are congruent, similar, or neither; much more interesting to me is the question of *in what context* we care about these different types of "sameness" for triangles. There are even contexts in which more things count as triangles. In abstract mathematics, we consider it acceptable for triangles to have one or more sides of length zero. Moreover, not only is it acceptable, but it's crucial for some constructions that those count as triangles. They are called "degenerate" triangles. So the shapes below actually count as triangles even though they look like a line and a dot, which I find quite satisfyingly subversive.

The first one is what happens if one side has length zero, so for example you could imagine the dotted edge here getting shorter and shorter until it's zero:

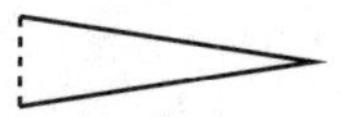

The second one has all three sides with length zero, so the whole thing has shrunk down to a point.

In my field of research, category theory, we call things a "triangle" if they have three sides, regardless of whether the sides are drawn with a straight edge or not. This is because in category theory we only care about relationships between things, and the shapes we draw are representing abstract relationships. For example, this relationship,

$$A \longrightarrow B$$

doesn't count as any different from this one,

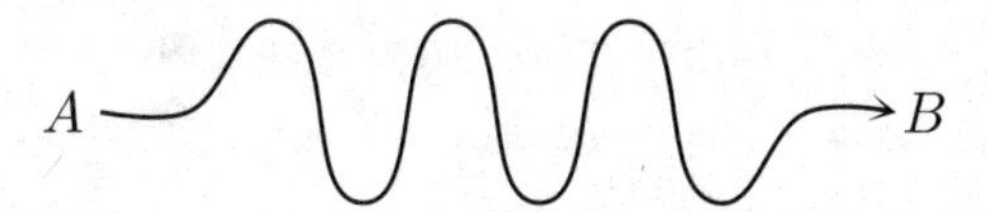

and these two things both count as "the same" triangle

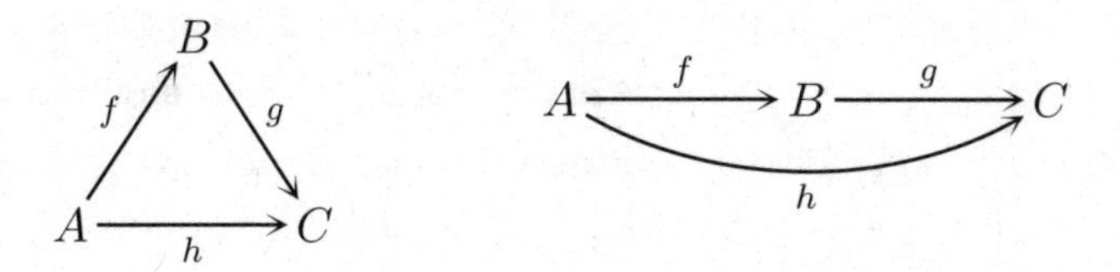

That might sound weird, but perhaps you'd be comfortable if I described the following scenario: when I was a grad student I basically lived my life going in a triangle between my room, my college, and the math department. Now, I've called it a "triangle" but of course I did not actually walk in a straight line between those places, because that's not how the streets were laid out. But it still felt like a triangle, although it was actually like this:†

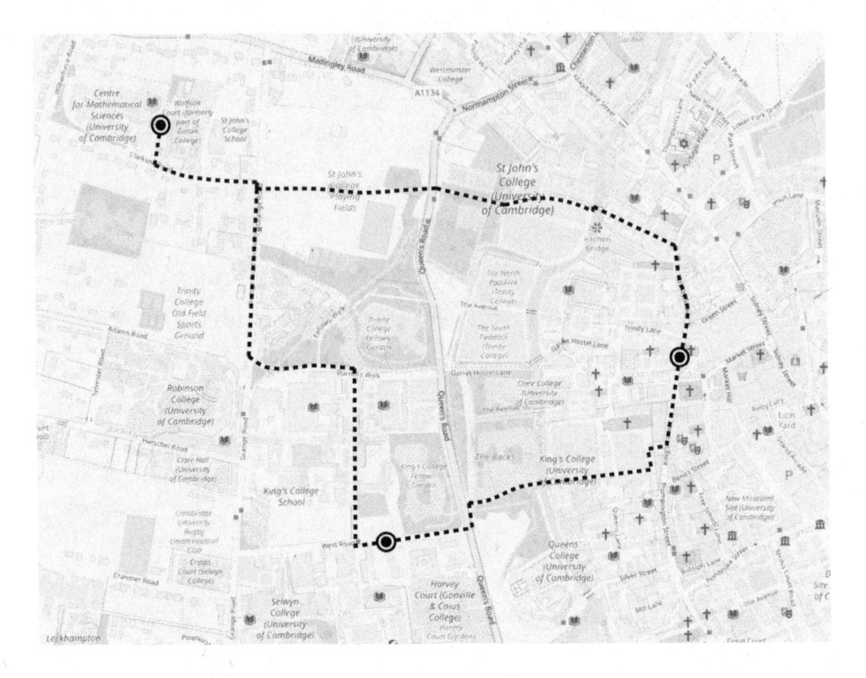

This discipline of finding senses in which things are the same and different is the starting point of all math, and it also comes in when we're thinking about when one plus one is and isn't two. We start by practicing this discipline with something fairly simple like triangles before moving on to things that are more complex. Unfortunately the simple things can seem pointless if nobody explains what we're trying to practice. Once we're well practiced, it becomes much easier to see connections in more complicated situations, such as the spread of viruses.

Viral illnesses spread by repeated multiplication. The idea is that each infected person infects a certain number of people, on average. Let's suppose that number is 3. Then those 3 people each infect 3

† Map image © OpenStreetMap contributors. Data is available under the Open Database License. See www.openstreetmap.org/copyright.

people (on average), which makes $3 \times 3 = 9$. Those 9 people infect 3 people each, which makes $3 \times 9 = 27$. At each stage the total number of new infections multiplies by 3.

Repeated multiplication is something that mathematicians study abstractly, just like they study repeated addition. Repeated multiplication produces exponentials, and this is a crucial point: if we say something "increases exponentially" in normal life, we might just mean it increases rapidly. But in math this means something very specific: it means it increases by repeated multiplication. This does indeed make for rapid increases, but rapid in a precise way that we can then study using other techniques. Among other things, this is how the spread of viruses in different conditions can be studied and predicted, even when it seems to be spreading very slowly at the beginning when the numbers are small. The graph of an exponential looks like this:

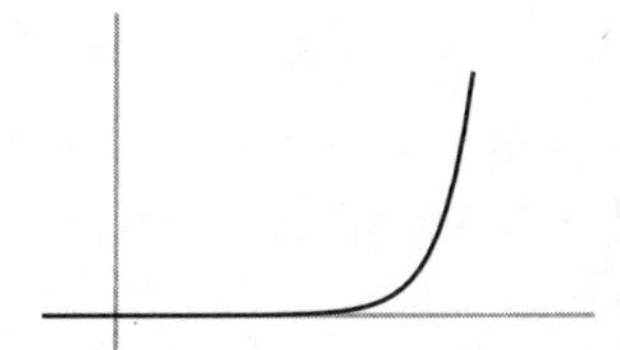

As you can see it's very flat at the beginning and then takes off rather dramatically later on. Studying exponentials abstractly enables scientists to have a better understanding of virus outbreaks even early on when the numbers don't seem too bad; unfortunately those who don't understand exponentials think that this is just fearmongering.

This turns out to be related to viral videos. When a video "goes viral" it's often something of a surprise and it happens rather suddenly. The point is that this can be modeled abstractly just like the spread of viral illnesses, but instead of infection, we're thinking about how the video is shared. So each person shares the video and that causes a certain number of their friends or followers to share it in turn. Even if that number is quite small, say 3, on average, then if the process keeps going the numbers will quickly become large because of how exponentials work. It only takes thirteen steps to get us past a million.

Exponentials govern all sorts of other situations that might otherwise seem unrelated, such as the temperature of a joint of meat as it's cooking. You can buy a meat thermometer gadget that not only takes the temperature of your meat as it's cooking, but connects to an app, which then predicts how much longer it needs to cook before it gets to your desired internal temperature. This is a calculation involving exponentials. Radioactive decay also involves exponentials, but this time it's something that is repeatedly multiplied by a number less than 1, so it keeps getting smaller.

There are some differences between these situations as well, besides the fact that some of them are more life-threatening than others. For both the viral illnesses and the viral videos, there is a constraint on the spread coming from the total population that is available to be infected (or affected). Once a certain proportion of the population is infected (or has seen the video), the spread will slow down even if we don't intervene to stop it, just because there aren't very many people left. This doesn't really happen when you're cooking meat—although I suppose if you leave it in there for a really long time it will eventually burn and disintegrate. It does happen any time the exponential growth is constrained by a limited resource, which is also the case with population growth, as a population will eventually start running out of food resources. The model of exponential growth limited by a constrained resource is more subtle than plain exponential growth. It was originally studied by Belgian mathematician Pierre François Verhulst in the mid-nineteenth century, and it produces a graph like this:

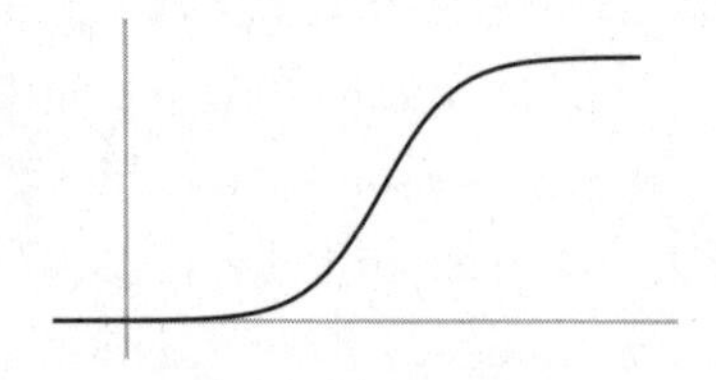

It looks a lot like an exponential at the beginning, but instead of growing forever it flattens out.

All this is to say that the math of these situations involves both finding similarities between the situations and also working out where the differences are, so that we don't try and push analogies further than is appropriate. This approach also helps us avoid pushing the concept of "correct answer" further than is appropriate. Instead, we find similarities between situations in which something is true, and use those as clues to help us understand what is making that thing true in those particular contexts. This is how we explore different possible answers to 1 + 1.

When 1 + 1 is not 2

Right at the beginning of this chapter we looked at examples of 1 + 1 being various different things. It's possible to dismiss those as "not really math" or "not really numbers" or "not really addition," but mathematicians prefer to study what's really going on in those places, partly to understand them better, and partly to get a deeper understanding of when 1 + 1 *is* 2. I remember when I learned to drive (on a manual car as is typical in the UK) my instructor had me stall the car deliberately so that I could get a better understanding of how the clutch worked. Exploring when something doesn't work gives us more understanding of when it does work.

Exploring when 1 + 1 makes different things also involved similarities and differences. We looked at situations in which 1 + 1 = 0, but then we also found some similarities and differences between those. Some of them were 0 because something "canceled out," like the nots, or the sign errors in the exam. Others were zero because the entire world was zero, like my zero world of sweets when I was little. Those are slightly different situations.

When the entire world is zero we have 1 = 0. This might look "wrong," but it's only wrong in ordinary worlds of numbers. It's correct in the zero world; it's correct in other worlds of more complicated objects with more complicated notions of 1 and 0 as well.

The worlds in which something cancels out are different because in those worlds 1 and 0 are genuinely different, it's just that 1 happens to cancel itself out. So these worlds can be encapsulated as an abstract structure with two objects and a way of combining them that satisfies this particular "canceling" relationship. We can represent that in this little table:

	0	1
0	0	1
1	1	0

If we interpret the 0 here to mean "0 negative signs" and 1 to mean "1 negative sign" then this table is telling us about multiplying positive and negative numbers, which has the same pattern as adding odd and even numbers:

×	positive	negative
positive	positive	negative
negative	negative	positive

+	even	odd
even	even	odd
odd	odd	even

Often in math we see a pattern arising in different places, we isolate the pattern (as in this grid) and then once we've noticed it, we start noticing it in other places that we hadn't previously thought of. I've used this particular pattern to help me understand tolerance. We can sometimes get ourselves tied up in knots if we're trying to be tolerant, wondering if that means we have to be tolerant of intolerant people. However, I think it's another of these same patterns as shown in the table below: I think that if we're tolerant of intolerance then we're allowing intolerance to flourish, so instead we should be intolerant of intolerance, and that this counts as tolerance.

	tolerant	intolerant
tolerant	tolerant	intolerant
intolerant	intolerant	tolerant

Those were all contexts in which things cancel out making 1 + 1 = 0. These are abstract structures that you wouldn't usually study until upper level undergraduate math classes. They are called "the cyclic group of order 2."

What about situations in which 1 + 1 = 1 because adding another 1 has no effect? These can be encapsulated as an abstract structure, which also has two objects (as in the previous patterns) but a different type of relationship. Here it's not that something cancels out, it's that something can keep piling up on itself without really having any further effect. We can depict that in a diagram like this:

	0	1
0	0	1
1	1	1

It's a bit like a diagram for dominant and recessive genes. In this case 1 is "dominant" and 0 is "recessive," so that in order to get a 0 as the result you have to start with two 0s, whereas if you just have one 1 that will ensure a 1 as the result.

The "tolerance" pattern and the "dominant–recessive" pattern are two different structures, but once we've made that observation we can think about the structures themselves as tidy little packages that we've already understood.

Packages

Wrapping things up in a package is a useful method for carrying more things at once. It would be rather hard to carry a dozen eggs if we didn't have them in an egg carton. Some things can be bundled up into quite general packaging where we just throw a bunch of things into a bag, but things like eggs benefit from something rather more careful and specific. It can then also be satisfying to discover that the specific packaging can be reappropriated for something else, such as a paint palette or for growing seedlings.

Making abstract connections between different things is also a way of packaging things up so that we can carry them around more efficiently, it's just that the things we're packaging up are now abstract, so "carrying them around" now means carrying them around as thoughts in our brain, rather than physically lugging them from place to place. Mathematicians can think of all the "tolerance–intolerance" type scenarios collectively as "the cyclic group of order 2" and this is now a single thought that we can carry around, leaving the rest of our brain free to do some more thinking. This is something that we even learn to do as children, for example when we learn to read.

When we first learn to read we start by recognizing individual letters. We then have to learn how to interpret those as words. When we're at the stage of having to recognize each letter one at a time it is extremely arduous to read whole sentences, so we then (consciously or otherwise) start recognizing entire words at once rather than letter by letter. This is really a way of packaging up a group of letters into a single unit so that our brain can carry it around more easily. It also means that our brain can "error-correct" or fill in the gaps, so that we can recognize words even if they are missing letters or have a typo.

We then take this up another level by starting to package words up as whole sentences instead of reading word by word; this is an aspect of speed-reading. This way of packaging things up more and more is also how we study pieces of music, especially long and complex ones. When you watch a pianist playing from sheet music, it might seem mysterious how they can possibly decode all those dots and squiggles so fast, but they're mostly doing it by packages. We don't read one note at a time, because, like with letters and words, it would be extremely arduous to read an entire piece note by note. Instead, we recognize groups of notes in chords, and we recognize groups of chords in chord progressions. If a piece is really long and complex we start packaging chord progressions into phrases, phrases into sections, and sections into movements. Once we've grouped a collection of phrases into a section, we can think of the section as a single unit and find relationships between whole sections of the music. In this way, a thirty-minute piece might end up just being five sections, which is much easier to think of than ten thousand notes.†

Math is about developing techniques for packaging ideas up so that we can get further with our limited brainpower. We've seen a couple of examples of this in repeated addition and repeated multiplication.

Mathematical packages

When we think about repeated addition we're thinking of things like 2 + 2 + 2 + 2. We could keep thinking of it like that but it would get quite tedious if we had a really long string of them. That's why we wrap that up in a new package called multiplication, in this case 4 × 2. It's not that important if we're not doing very many of these things, just like I can get two eggs out of the fridge and carry them to the counter just fine without any sort of package, but if I'm making a cake with six eggs I'll probably get the whole egg carton, and then put it back when I've cracked my six eggs.

It's not that hard to write out 4 × 2 as repeated addition, but it would be terribly tedious (and not very illuminating) to try and write out 44 × 22 like that. Multiplication is harder than addition, because it is a package of things, but if we can understand it as a single package then we can get a lot further with our thoughts. For example, we can then do repeated multiplication, things like 3 × 3 × 3 × 3. As with

† I just tried roughly estimating the number of notes in Beethoven's *Pathétique* Sonata and came up with around ten thousand, though it's only a twenty-minute piece.

repeated addition, it can help us to have a way of putting that in a single package, and in this case we write it with an "exponent," that is 3^4. It's a small act (like wrapping up a present) but it is a starting point that enables us eventually to take flight into the world of exponentials and understand things like viruses spreading.

This idea of packaging things up into units can then be used for things that aren't numbers, just like we tried adding and multiplying things that aren't numbers. We can package up the logic of our arguments (and other people's arguments) in this way as well, so that we can understand much more complex arguments, just like how we can read complex books or play complex pieces of music.

Logical arguments are built from "if...then" statements, called logical implications. We say "if this one thing is true, then this other thing must be true" and we progress from a starting point to a conclusion by piling these logical statements on top of one another, or rather, by lining them up end to end so that we can travel along them without leaving any gaps. Small children are less good at following several steps of logic or causation in this way. So for example they might get as far as thinking "If I stay up longer, I can spend more time playing with this toy now," but they won't get as far as thinking "but also then I won't get enough sleep and I will be cranky in the morning." To be honest, as an adult, I am often not very good at that line of argument either, preferring to stay up doing something fun. However, to give me my logical due, it's not that I haven't followed the extra piece of logic, it's that I've followed it and decided that on balance the painful morning is going to be worth it because of the extra fun now.

Playing chess famously involves thinking ahead to see what the consequences of each of your moves could be. Novice players might only think about trying to capture an opponent's piece right now, without noticing the weak position it leaves them in afterward; conversely, a more advanced player will knowingly sacrifice a piece now if it puts them in a stronger position later. I admit I never got much beyond the immediate aspects of chess myself, but I'm very good at building up and following complex logical arguments, which just shows we can be good at (and enjoy) these sorts of processes in some contexts even if we aren't good in others.

Packaging logical arguments into units can help us recognize patterns; this is true of logical fallacies as well as good logical arguments. A "straw person" fallacy (usually known as a straw man, but I try not to gender these things) is when someone replaces your argument with a much weaker one (a straw person) and then attacks that. For example, some people reject the concept of white privilege on the grounds that rich Black people exist. However, they're not arguing against white privilege, they're arguing against the idea that all Black people are poor, which is not an argument anyone is making: it's a straw person. Once you've understood that sequence of thoughts as a single unit (a straw person fallacy) it's much easier to think about what is going on and recognize it elsewhere, such as when people object to monuments to historical slave traders being taken down by claiming that we should not be "erasing history." Nobody is trying to erase history—that's a straw person. Removing a monument is not the same as erasing history; the history is still there, it's just a question of whether we celebrate it or not.

In fact, a straw person fallacy always has some sort of false equivalence inside it, because it's based on taking someone's actual argument and falsely equating it to a much weaker one, before knocking it down. What we've done now is not just package up the arguments, but we've also examined what smaller units we can break them into. Perhaps this is like using small packing cubes inside your suitcase, which is something I was very averse to doing until suddenly I was completely converted to it.

Packaging arguments up to understand them as units is helpful so that we can understand how they fit into broader contexts. There's also a sort of reverse process: breaking arguments down into constituent parts, and this is helpful so that we can understand exactly what is driving them and where they are coming from at root.

Building blocks

Understanding the basic building blocks of an argument is important if we care about understanding why other people think the things they do, from *their* point of view. There are always reasons, even if they don't seem logical to us, and if we are going to be empathetic humans it's important for us to find those reasons and acknowledge them. This is the principle of being able to break things down, or distill them down, to their basic building blocks.

Where human beliefs are concerned, the basic building blocks are people's personal basic principles or basic beliefs. Understanding them can help us understand the roots of disagreements between people, which often come down to a disagreement about some very basic principles, not, as people too often like to believe, that one party is being logical and the other party is not.

In a way, all of life is about understanding things by breaking them down and building them up, and the more techniques we have for doing that, the more things we will be able to understand. We should acknowledge that things are difficult, so we need to break them down into more basic parts to understand them. We also need to understand how to start with basic parts and put them together to build up those hard things. It's like making a tiered cake, where you first make the individual cakes, and then make the icing, and then start building it up. And it's important that the cake has enough strength to hold up the tiers, or we'll need to add in some structural pieces or some sort of cake stand to hold it all up.

In more formal mathematics, we have basic *axioms* for a mathematical world, which are the basic facts we assume to be true, before seeing what other truths can be built from the basic ones in that particular world. It's important that we're not saying those basic axioms are definitely true, we're saying that we're going to study a context in which those axioms are true, to see what else follows from it. This is something we can also do to understand someone else's beliefs:

identify their basic principles, and then see what else follows from them, without us having to agree or believe in any of the same things. It's all about studying different worlds in which different things hold, just like 1 + 1 is different things in different worlds.

So 1 + 1 = 2 might be a basic axiom in one world (the world of ordinary numbers), but in another world 1 + 1 = 0 is a basic axiom (in the cyclic group of order 2), and in another world 1 + 1 = 1 is a basic axiom (in the dominant-recessive world). So the question is no longer "Why does 1 + 1 = 2?" but rather "*Where* does 1 + 1 = 2?" and then "What else must be true in a world where 1 + 1 = 2?" or, more fundamentally, "What is a world in which 1 + 1 = 2?"

When 1 + 1 is 2

We've finally arrived at the abstract mathematical version of "Why does 1 + 1 = 2?" The answer, really, is that 1 + 1 doesn't always equal 2, as it depends what sort of context we're in. We can explore this context by thinking about its basic building blocks. We start with the idea of 1, and the idea of putting one thing together with another thing. Then we specify that this definitely makes two things, not zero things, not one thing, and not three things. We then ask ourselves what context this gives, that is, what else must be true in a world with those starting points?

This is essentially how we get "ordinary numbers," that is, the whole numbers, the counting numbers: 1, 2, 3, 4, . . . and so on. These are sometimes called the *natural numbers* (although those might also include 0, which is a whole different story). We build this world from 1 and a process of addition with no collapsing, disappearing or reproducing, and in abstract mathematics this is called generating a structure freely. The "free" part means we don't impose any extra rules on the situation, other than our basic axioms—we just let the thing grow organically and then we tiptoe in to see what sort of jungle has resulted.

So 1 + 1 doesn't equal 2 in *every* context, but there is a wide range of places in which it does, and these are encapsulated in the abstract world of the natural numbers. It is helpful for us to understand that abstract world, so that we can then look for places where 1 + 1 = 2 in the concrete world, and know that everything we've learned about the abstract world in which 1 + 1 = 2 will then be true in the corresponding parts of the concrete world.

There are many things to explore about this world in which 1 + 1 = 2. One thing we might wonder is: If we can add things together, can we take things away? That's a whole different level of question, and takes us into the mysterious world of negative numbers, and the next chapter.

CHAPTER 2

HOW MATH WORKS

Why does $-(-1) = 1$?

This is one of those baffling "facts" that some people think is obvious, and some people think is very mysterious. Is it just a "fact" that we need to accept and memorize? In fact, are there any mathematical facts that we need to accept and memorize? What is a fact anyway?

Some people immediately accept that $-(-1)$ is 1, and it bothers me that they are likely to be deemed to be "math people" and those who question it are deemed "non-math people." Well, it bothers me that those labels exist at all, making it seem that mathematical ability is innate and some people simply don't have it, whereas really everyone has mathematical ability of some sort and everyone can get better with the right sort of help.

Those who accept that $-(-1)$ is 1 are just people who accept that $-(-1)$ is 1, and those who question it are people who question it. Importantly, it's not a dichotomy: it's possible to accept it and also question it. Math is about questioning things and wondering, deeply, why things are true. If you think that $-(-1)$ being 1 is mysterious, rather than obvious, that doesn't make you "bad at math": it could mean that you are thinking like a mathematician. Wondering about things that might seem obvious to others is how a lot of deep math is developed. In this chapter we're going to look at how much abstract thought process it takes to arrive rigorously at this supposedly obvious equation. However, my purpose isn't really to explain the equation (though that's a side effect), but to explain how we decide what is true in math. In the process of investigating why that particular thing is true, we are going to investigate how we ever know that anything in math is right. In fact, how do we know that anything is right at all, ever? This chapter is about the frameworks we use for deciding what we will accept as true in math.

Everyone has different tolerance levels for accepting things as true. Some people will accept things as true if they read one article saying so on the internet, regardless of who wrote the article, how well sourced and cited it is, and whether any other articles agree with it. Other people are happy to accept something as true if it is declared true by someone they believe in, perhaps a senior professor, a trusted news source, a religious leader, or a political idol. Some people will believe something when it feels true to them, like that horoscopes are accurate or that homeopathy works or, in my case, that listening to Bach helps me do better mathematics.

Academic subjects all have a framework for assessing the truth of things, a framework that is trying to be better than "this feels true" or "I said it therefore it's true" or "I read it on the internet so it must be true." This is to try and get our understanding of the world around us onto a more secure footing than just random opinions or guesses that won't hold up to further scrutiny. It's to do with wanting to build on our understanding, not just stare at it, like the fact that if we want to build a tall building we'll need better foundations than if we're just pitching a single-person tent. Why anyone has this urge to build tall buildings, metaphorical or otherwise, is something we'll come back to when we think about some uncomfortable connections between academic research and colonialism.

Uncomfortable connections notwithstanding, academic disciplines are all built from the starting point that some framework is desirable. The framework is there so that we can reach some consensus about what counts as good information, and then build on it in ways appropriate to that framework. It's a bit like agreeing on rules for a particular sport, and then building up teams and tournaments and championships inside that framework. It doesn't mean the outcomes in those tournaments are "correct," it just means those are the outcomes determined by that framework.

Moreover, the frameworks should be objective, and not rely on just believing one authority figure. However, they can end up looking like they depend on authority because one aspect of a framework is that it identifies "experts" who have been verified as expert according to the framework. So these experts are not ruling by random authority: they have been established as proficient by the framework, and in principle anyone can become more expert by becoming more proficient according to that framework.

The framework of mathematics is logic, and the reason I am drawn to it is that I don't want to have to trust other people to determine what information counts as true. I don't want to have to trust books either, but understanding the framework of logic makes me better able to decide which books to trust more than others, and indeed which articles to trust more than others even if they are online. Some people too easily believe what they read online, and there's a backlash against that, saying that we shouldn't trust biased news sources and we shouldn't trust Wikipedia (for example). A more productive position is that we should learn how to assess things, so that we don't have to trust those sources, but we also don't have to just dismiss them out of hand.

How do we know math is right?

This chapter is about how math works, and we are thinking of this in a particular sense: What is the framework for deciding things are right in mathematics? Different academic subjects have different frameworks for deciding what counts as good information. Science uses evidence, and has clear frameworks for what counts as good evidence.

Importantly, evidence-based scientific results aren't absolutely true, they are scientifically true, that is, they are supported by a certain scientific framework. This often means that there has been a certain level of testing and the evidence supports the conclusion to a certain degree of certainty, perhaps 95 percent, or 99 percent depending on how critical the situation is. This might sound like science doesn't know anything for sure, because it doesn't have 100 percent certainty. That is sort of accurate, and it's dangerous for us to claim that science is absolutely right when it has this uncertainty built in. It's much better for us to understand what that uncertainty means than pretend it isn't there. We can then understand that "not knowing for sure" doesn't mean that everything is equally likely. If scientists say they are 95 percent certain that most of modern global warming is caused by humans, then it's a whole lot more likely to be true than not.

Math doesn't work by evidence: it works by logic. Logic is how we decide that something counts as "right" in math. This doesn't mean it is right; it means it is right according to the framework of math, which is logic.

This brings us to the important question (or possibly complaint) of why, in math, you have to "show your work." This is the bane of many children's math existence: they have some math questions to answer, they know the answers, they write down the answers. The answers are right, but they didn't "show their work" and so they don't get full credit.

Is that fair?

The point is that math isn't just about knowing the right answers to things, but too often it is presented as if it is only about getting the right answers to things. It is presented as if there are facts that you have to know. The facts are announced by the teacher, and the students' role is to learn those facts, not question them. As a result, the role of the teacher is to announce those facts, not justify or explain them. This is an extreme example, and I'm not saying that all math teaching is like this, but too much of it is too much like this.

This gives students the impression that math is based on authority, that there is a truth imposed from above, like a decree from an autocratic power, and we humans just have to obey the decrees without question. This is not only an incorrect portrayal of what math is, but it's also a dangerous attitude to pass on to children.† If they think that knowledge comes from authority, then they risk becoming adults who get their knowledge from authority figures rather than from objective frameworks. The result is that what they believe depends on who they see as an authority figure, and then we can't reason with them because their beliefs aren't based on reason; they're based on authority.

This is almost the exact opposite of what math really is. The whole point of math is to deduce things by logic. Then nothing needs to be handed down by authority except the foundations of logic itself. The trouble is that school math often consists of questions with answers, and an answer key telling you what the answers are. Then you can check to see if you have "the right answer" by comparing it with the answer key.

But research math has no answer key because we don't know what the answers are yet. Life definitely has no answer key. So the question is: If there is no answer key, how do we decide if we have a good answer? That is the point of math: learning how to decide what counts as a good answer when there is no answer key. And that's why it's crucial to show our work: because the work is the math. Math is not about "getting the right answer," it's about building arguments backing up an answer.

We are going to investigate the idea of $-(-1)$ by asking a lot of questions about what all these things really mean. This is a typical way that mathematicians dig into their intuition to unpack it and understand it logically. Asking searching questions leads us to a much stronger

† I found this particularly vividly expressed by Professor Dave Kung in his TEDx talk "Math for Informed Citizens," at https://youtu.be/Nel5PF8jtsM.

logical footing, as long as the purpose of our questions is to understand things more, not just to thrash out and topple them.

Imagine if we were designing a jungle gym for children. We'd want to test it in every possible way to make sure it's safe. We wouldn't test it by just playing on it in sensible ways: we'd want to jump on it, swing from it, bash into it, fall from it, and try and pull it out of the ground, rather than simply trust that we built it well. The solidity of math comes from not wanting to trust things, but wanting to jump and swing and know that our framework will hold up. One of the reasons the framework is so strong is precisely because we question it so deeply. The simple-sounding questions we ask aren't merely "not stupid," they're of critical importance.

I will admit that digging into some of these questions can sometimes seem a bit arduous. In normal life we accept a certain quantity of things as true so that we can get on with our lives. When toddlers question everything with unending curiosity it can be a wonderful opportunity for exploring the world, but also sometimes you really need them to get their shoes on so that you can get out of the house.

Then again, in math we're not trying to "get on with our lives"—we're trying to establish strong structures. It might seem arduous to make a strong structure for a house, but that's better than skipping steps in order to get the job done more quickly. That said, when we're doing research in math we're often rather more impressionistic at first, in order to try and work out where we're going before the images in our head dissipate. But that's more like sketching out the first ideas for the design of a building, before sitting down and actually building it.

Understanding why $-(-1) = 1$ is going to involve thinking really hard about negative numbers and what they mean, and this is in turn going to involve thinking hard about zero and what that means, which will then lead to thinking about numbers and what they are in the first place. In thinking about those things we'll see how mathematicians developed better ways of reasoning, often when they realized they had been taking too much for granted about what things really mean.

The idea of negatives

Negative numbers are difficult. Positive numbers are already difficult. We've seen that they come from us noticing an analogy between different collections of objects, which we then turn into an abstract concept. But negative numbers can't come from an analogy between concrete things, because we can't really see "negative things." We can't count −2 strawberries and −2 bananas and say "Aha! The thing that these collections of objects have in common is the concept of −2."

Instead, there are some other typical ways of trying to get our heads around the concept of negatives. One way is by thinking about changing direction. If you walk ten paces forwards, and then ten paces backwards, you'll get back to where you started. Those directions cancel each other out, and we can call the backwards direction "negative." This is a decent intuition, but is not exactly logical (though it's also not illogical), nor is it very transferable. How would this help us understand negative ten of anything else, like negative ten apples or negative ten dollars?

Another way to try and understand this is via debt. If you owe someone ten dollars it's not just that you don't have ten dollars, it's worse than that, because you're actually down ten dollars. But that's already an abstract concept, and for children who've never owed anyone anything that's a hard concept to get your head around. You either have some cookies or you don't have some cookies. What does it mean to owe your friend a cookie?

Well, what it really means is that you need to get a cookie from somewhere and give it to your friend. Normally if someone gives you a cookie you have a cookie. But owing your friend a cookie means that if someone gives you a cookie you are morally obliged to give the cookie to your friend, and the end result is that you have zero cookies.

It is a bit odd to bring moral obligation into a discussion of numbers, and all this is pointing to the fact that negative numbers really are a difficult concept, and we should acknowledge that. They are a level more abstract than positive whole numbers, because at least positive whole numbers are an abstraction from something concrete (objects). However, negative numbers are an abstraction from something that was already abstract. I fear you might feel like throwing up your hands and giving up at this point, but I rather like the fact that our brains are able to do this. I like the fact that, as we grow up, we become increasingly able to deal with hypotheticals, as long as we are comfortable using our imagination and wandering around in an imagined world. It's a bit like magic realism, which is, in a way, a level beyond fiction. Fiction is an imagined scenario based in the real world, but magic realism is an imagined scenario based in an imagined version of the real world in which slightly different things are possible. Fantasy is, perhaps, even further than that: an imagined scenario based in an imagined version of the real world, which perhaps doesn't even start from the real world, but from something completely different.

I find it intriguing when our suspension of disbelief can only go so far. There are some books (which I'd better not name as this would constitute a large spoiler) in which the entire book turns out to be a fiction invented by one of the characters in the book. Some people are upset by this big reveal, and criticize it by saying it's implausible that someone would make up an entire fiction like that. This is despite the fact that the book is, in reality, written by an author: those readers are apparently able to accept that a real human might write a fictional book-length story, but not that a fictional human might write a fictional book-length story inside a real book.

Everyone has a different tolerance for fantasy in fiction, and a different tolerance for abstraction in math. Some people only like reading nonfiction; some people like fiction but not magic realism. I personally like magic realism but not fantasy. However, I love abstraction in math: I don't just tolerate it, but enjoy and appreciate it. I enjoy it in its own right, but I also appreciate what it's trying to do. Abstract mathematics consists of dreaming about imagined or hypothetical versions of reality, and sometimes piling up further and further levels of hypothetical, but with a purpose: the aim is to shed light on reality. Negative numbers are, in a way, a fiction made up by mathematicians to encapsulate various scenarios in life that aren't quite to do with straightforward counting. The scenario of walking forwards and backwards might not seem like debt, but if we think about it a little more abstractly perhaps we can see what the connection is, in the same way that we were finding connections between things in the previous chapter.

In the forwards-and-backwards scenario we said that if we walk forwards ten paces and then backwards ten paces we get back to where we started. In the debt scenario we said that if we owe our friend ten cookies then if someone gives us ten cookies the end result is we have no cookies, because we have to give those ten to our friend. Both of these scenarios use the idea of returning to nothing. And so, in order to understand negatives we need to understand zero.

Zero

Zero is a baffling number because it represents nothing and yet is still a thing. It is a thing representing nothing. It is difficult to look at a collection of zero strawberries and a collection of zero bananas and observe what they have in common, because it is hard to see zero things. I was once delighted to discover that I owned three three-hole hole-punches, two two-hole hole-punches, and one one-hole hole-punch. Some people declared that I also owned zero zero-hole hole-punches but I wasn't so sure. Perhaps everything that doesn't punch a hole is a zero-hole hole-punch? In that case my water bottle is a zero-hole hole-punch, and so is my computer, and all my coffee cups, in fact almost everything I own is a zero-hole hole-punch.

Similarly, almost everywhere we look we can "see" zero strawberries, and also zero of a whole load of other things. This is rather confusing; if you're anything like me you're now getting a bit dizzy realizing that everywhere you look you can see zero of an infinite variety of things. I need to blink a little and take a few breaths to come back to reality. In fact, that sense of suddenly "seeing" all those non-things in your imagination is definitely a typical feeling I get when doing math. It comes with a slight sense of vertigo, the feeling of an infinite imaginary world suddenly flashing before my eyes, combined with bafflement, confusion, excitement, and then coming back to reality. I enjoy it.

The concept of zero has a long and varied history, and could fill an entire history of math book. That's not my purpose here; my purpose is really to validate the feeling that zero is a bit weird. There's also a difference between the concept of zero and the decision to consider that concept as an actual number and include it in a number system. That's a subtle but important distinction, and for me that's related to the question of whether math is invented or discovered. My stance is that the concepts exist already and thus are things we discover, but our methods for writing them down and reasoning with them are human constructs and thus are things that we invent. Sometimes the concept and the way of studying the concept aren't so easy to distinguish, and so I don't find it possible to say whether it was invented or discovered (nor do I find it important to do so).

Many different ancient cultures considered the concept of zero and used a symbol for it, including Egyptian, Mayan, Babylonian and Indian cultures. The ancient Greeks were more worried about its status and not convinced it should count as a number.† To this day different people have different levels of comfort about what gets to count as a number. Most people these days seem to be fine with zero and with negative numbers and fractions, but when we get to more advanced concepts like "imaginary numbers" it's trickier. (Perhaps you don't know what those are, but we'll come back to them in Chapter 4.) I have received angry emails from people telling me imaginary numbers

† Note that some of the people we usually call "ancient Greeks" were from other parts of the Greek Empire, not actually Greece. I'll come back to this in Chapter 4.

shouldn't be called numbers because they're not numbers (not that it's my fault they're called imaginary numbers, but that doesn't stop the letter-writers). Really, wondering what counts as a "number" forces us to ask what numbers are at all, and it reminds me that across history we have become more and more accepting of what counts as a number, even if some people lag behind. This is not surprising: we have become gradually more and more accepting in society too, but some people lag behind, perhaps accepting women and Black people but not gay people, or maybe accepting gay, lesbian and bisexual people but not transgender people.

One way of taking a shortcut through the difficult question of whether or not zero is a number is to deftly sidestep it. Does it matter what a "number" is? Instead of worrying about that, we can simply study a world in which zero is a basic building block, and see what sort of world that is. In that system, we don't have to say what zero is or what it represents, we just have to say how it interacts with everything else in the system.

The usual way of doing this in math is to build up from a previous world. So far we have built ourselves a world starting with the number 1 and building by addition. This got us all the numbers 1, 2, 3, . . . and so on. When we include 0 in this world we need to know what happens if we "build" it onto other numbers, that is, what happens if we add it to other numbers. Secretly we know we're trying to represent nothing, so we can declare that adding 0 onto anything else doesn't change it. So $1 + 0 = 1$, and $2 + 0 = 2$, and $3 + 0 = 3$ and so on. We can't list all those equations, as there are infinitely many of them, so we can instead encapsulate this thought:

> If we start with any number and add 0
> the answer is the same number we started with.

That's a bit long-winded, so we can give a name to the number that we start with: we could call it x and let that stand for "any number we start with." I have now turned a number into a letter and this might make you shudder, and we'll come back to that idea later. But for now I hope you can at least see (I mean literally, on the page) that doing this enables us to rewrite the above statement to be a whole lot shorter, as follows:

For any number x, $x + 0 = x$.

This might seem a bit like a cheat because we haven't actually explained in what way zero is a number representing nothingness; instead we've characterized it by what happens when you add it onto things. This is typical of abstract math and could be seen as pragmatic rather than illuminating. We've turned something instinctive into something that we can reason with. The result is that we can reason with it, but we might have less intuition about it. The tension between those things is ever-present in abstract mathematics.†

Anyway, this is how we get ourselves a world including zero: we just put it in. We say "I declare that there is something called 0 and it behaves according to the above statement," and then we start playing around with it. We can then do something similar to get ourselves negative numbers.

Negative numbers

To build a world with negative numbers we will do a similar sidestepping trick to the one we did for zero: we don't say what negative numbers are, but instead we say what they do. And it's just like "going ten steps backwards" to get back to where we started. The idea of a negative number is that it's some way to "get back to where we started." In this case, the place we start is zero, which is why we needed to have a concept of zero in our world.

† David Bessis goes into this in some depth in *Mathematica*.

So we decide to include some new building blocks in our world, specifically for the purpose of getting back to where we started. The definition of −1 is "something that cancels out 1." It's like antimatter. When I was little I thought that pepper was antimatter for salt, that is, I thought that if you added too much salt to something you could add some pepper to cancel it out. I'm still a bit sad that this isn't the case, and that it turns out there's not much you can do to fix it if something is too salty; also I don't like pepper, so pepper is a double disappointment to my adult self.

A more formal way of saying that −1 cancels out 1 is to say that we add it to 1 and we get 0. This concept of something that cancels out something else is called an inverse, and more specifically in this case it's an additive inverse because we're undoing a process of addition. Now, we don't just want to be able to cancel out 1, we want to be able to cancel out any number.

At this point I realize you personally might not want to cancel out anything at all, but I'm trying to explain the mathematical urge behind all of this. So when I say "we want" what I really mean is "this is the mathematical urge." I know we all have different urges. Some people see an open cupboard door and feel the urge to close it—I don't! Some people see a mountain and really want to climb it—I don't with that one either, but I do feel mathematical urges. They are sometimes urges of extrapolation: I canceled out one thing, now I want to see if I can cancel out everything else. I do get those urges in the kitchen, say, I've tried making cake with one kind of flour, now I want to see what happens if I make cake with every kind of flour: wheat flour, oat flour, almond flour, rice flour, coconut flour...

When we follow this particular mathematical urge, we find that once we've given ourselves −1 as a building block for canceling out 1, we can use it to build everything we need to cancel out every other whole number. This is because all the whole numbers are built by sticking 1 together a bunch of times, so we can use −1 the same number of times to cancel those out.

For example, if we stick two of the negative ones together, that will cancel out two of the positive ones. If we write this out in symbols it says this:

$$(-1) + (-1) = -2$$

This might look like the definition of -2, but that's not quite what it is: the definition of -2 in this point of view is "a thing that cancels out 2."†
We have the following steps:

- We have a building block 1.
- By definition in this world, 2 is built as $1 + 1$.
- By definition in this world, -2 is whatever cancels out 2.
- $\{1 + 1\} + \{(-1) + (-1)\} = 0$, so $(-1) + (-1)$ cancels out 2.
- So $(-1) + (-1)$ is -2.

If we now remain calm we can figure out what $-(-1)$ is. Remember that negatives are things that cancel out other things. That's rather vague, so I'm going to use a letter again, to represent a "thing": $-x$ means "the thing that cancels out x (by addition)."

So "$-(-1)$" means "the thing that cancels out (-1)." Well, the thing that cancels out (-1) is 1, and this is why $-(-1) = 1$. We could write it out in steps as above:

- We have a building block 1.
- -1 is whatever cancels out 1, meaning $1 + (-1) = 0$.
- $-(-1)$ is whatever cancels out -1.
- But the equation $1 + (-1) = 0$ also tells us that 1 cancels out -1.
- So 1 is $-(-1)$.

† There is a slight subtlety which is that we also have to check that for any x there can only be one thing that cancels out x, otherwise this definition of $-x$ would be ambiguous.

That might have seemed arduous or it might have seemed revelatory. I've been interested to note that people who have been deemed "good at math" in school might well find it arduous, whereas people who have been deemed "bad at math" in school might find it revelatory (like my art students often do). I can't remember how I felt when I first learned about it, but I do know that it remains the only satisfying explanation to me, because it really digs into the roots and meanings of things.

You might wonder why we should care. In fact I don't think anyone *should* care about all that, because we all care about different things, and really the only thing I want everyone to care about is reducing human suffering, violence, hunger, prejudice, exclusion and heartbreak. After that, I hope that everyone will care to think ever more deeply about why and how we count things as true or not.

We run into problems in the world if people assume they're right with no framework for assessing their rightness. We end up with contradictions, disagreements, conspiracy theories. It's a slightly tricky balance because we mix up opinion with fact in both directions: sometimes all opinions are equally valid but people claim they aren't, and sometimes not all opinions are equally valid but people claim they are.

Some things genuinely are just an opinion and everyone is entitled to a different one, such as when it comes to personal taste, in food, or music, or films. However, sometimes people think there is a right and wrong about those tastes. The fact that I'm not a huge fan of toast or Mozart doesn't mean I'm wrong, because there's nothing to be wrong about—I just don't like it. (And yet, people try and tell me I'm wrong about this all the time.)

However, there are other situations where not all opinions are equally valid. If something has a huge quantity of evidence backing it up, then I think it is much more valid to believe that than to believe something that has essentially no evidence backing it up, like that the earth is flat or that the 2020 US presidential election was stolen by the

Democrats by means of widespread fraud. (There is no evidence of widespread fraud, but by contrast plenty of evidence of gerrymandering and voter suppression in favor of Republicans. This doesn't mean that one side is necessarily right and the other wrong, but it does mean that one side is going against evidence.)

Mathematicians specifically don't assume they're right themselves, no matter how right something may feel. The feeling is often the starting point, but the feeling can lead us astray, so we keep questioning it to make sure, and we don't count it as right until we have a rigorous logical argument backing it up. This self-questioning is not always good for our self-esteem, but it is often how the foundations of mathematics are made more secure, and this in turn enables more mathematics to be built up. This even happened relatively recently (in the context of the history of math) when mathematicians tried to pin down what numbers actually are.

When mathematicians get insecure

What are numbers? We sidestepped that question earlier but it gets harder to sidestep it if we try and look at more complicated numbers. It took quite an elaborate process to come up with the world of whole numbers including negatives, but we were able to do it. It's a step more elaborate to come up with the world of fractions, which are also called rational numbers as they represent ratios. But it's a markedly different matter if we try and think about so-called irrational numbers. For these, it's actually not so hard to say what they're describing in the world, but it's very hard to build them as a mathematical world. When we did that for the whole numbers we started with the number 1 as a basic building block and took it from there. For irrational numbers it's not at all clear what the basic building blocks should be in the first place.

At this point you might not remember (or have ever known) what an irrational number is, and so I should say what they are. The trouble is that it's extremely difficult to say what they are. They are sometimes thought of as "decimals that go on forever without repeating," but what on earth does that mean? If a decimal goes on forever without repeating, how do we know what it is? No matter how far we list the decimal places we'll have left some out (infinitely many, actually), and we can't describe them by a pattern because the whole point is that they never repeat, so there is no pattern we can use to describe them.

This is another of those places where if you feel weird and vertiginous and dizzy and overwhelmed those are good mathematical instincts. Too often it can seem that people who are "good" at math take all these things in their stride, and so anyone who is baffled by them duly feels like "not a math person." This is the wrong way around. If someone is taking all that in their stride then there are some subtleties they are overlooking.

The story of irrational numbers is a long one. The idea of them arose a very long time before mathematicians worked out a way of understanding them. The mathematicians of ancient Greece had already worked out that there are some things that can't be described as a fraction. It's not that hard to come up with such a "number" (though it is a little harder to prove it's not a fraction), for example we can just consider this square:

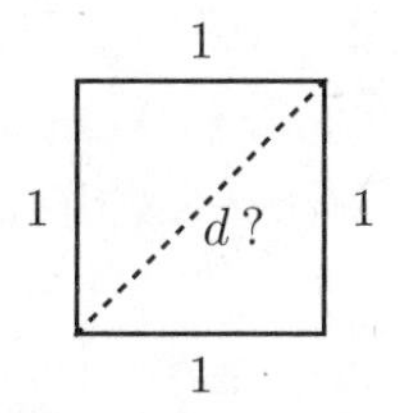

This square has all its sides of length 1. You might wonder what units I'm using, but the glorious thing about abstract math is that it doesn't matter. It's 1 of whatever I feel like, and because it doesn't matter, I don't need to specify it.

Now we might start wondering how long the square's diagonal is. (As before, you might not start wondering that at all, but what I mean is that it's a mathematical urge.) If you remember a little Pythagoras you can work it out: Pythagoras's theorem says that for a right-angled triangle, "the sum of the squares on the shorter two sides equals the square on the longest side." This is expressed a little more succinctly as follows (using letters!):

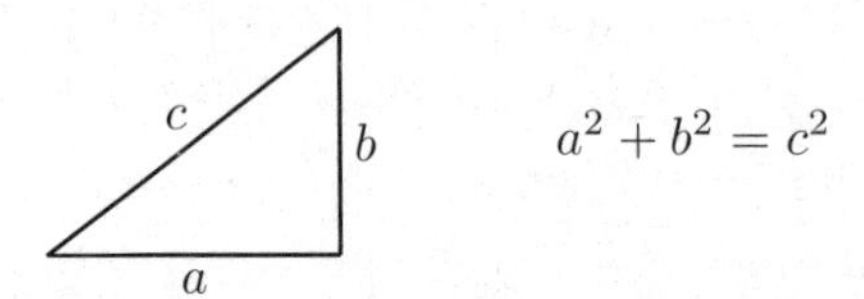

We can now apply this to one of the triangles in the square we were trying to understand. If the length of the diagonal is called *d* (we've used a letter again) then the formula gives us this:

$$1^2 + 1^2 = d^2$$

which tells us

$$2 = d^2$$

This is saying that *d* is a number with the property that when you square it the answer is 2. However, it is possible to prove that there is no fraction *d* that can possibly satisfy this property. That means we have two choices: either the diagonal of that square is not a measurable length, or there must be some lengths that are measured by numbers that are not fractions.

The first option is logically not disastrous, it's just rather limiting and unsatisfactory. How can there be lines that have no length? It's a good mathematical instinct to feel that that is strange. If we only allow fractions to be numbers, then there will be tiny gaps in between them where there is no number. We could put another fraction inside the gap, but there will always be a gap if we zoom in enough, like the fact that if we zoom in on a curve on a computer screen then eventually it will turn into individual pixels.

If there are tiny gaps in between the numbers this would also mean that, say, when we grow up there are some instants at which we have no height. That really is quite a bizarre idea. The alternative is to allow some new types of number into your life.

You can always be determined not to let more things into your life, and many people do steadfastly refuse to accept new ways of being as valid (such as same-sex marriage, or nonbinary gender, or women mathematicians) but that's not how mathematics operates.† Contrary to the idea that math is fixed and rigid, mathematics always wants to let more things in. It doesn't necessarily want to let more things into any particular world, but it always wants to investigate a new world in which those things can happily coexist with the ones we had earlier.

So math takes the inclusive option for the diagonal of that square, and, rather than declare that it has no length, we acknowledge that there must be some numbers that are not fractions. Then the question is: If they're not fractions, what on earth are they?

In 1872, mathematicians Georg Cantor and Richard Dedekind were (separately) worrying about how to teach numbers to their students rigorously. I like to imagine that, as good teachers do, they were preparing their lectures by imagining their students' reactions, empathizing with them, and preempting the sorts of questions their students would then ask. This pushes good teachers to understand things more deeply, because in order to explain ideas to students coming from many different points of view, you have to understand the ideas from many points of view yourself. Cantor and Dedekind both realized that mathematicians, collectively, had not set up number systems at all rigorously, so they set about doing it themselves.

It is one of those curious features of humanity that they both did this at around the same time, but quite differently. Both of their ideas are rather difficult to explain without going through a lot of technical

† Unfortunately, it is how some individual mathematicians operate.

background, but I'll try to hint at the ideas. Cantor's idea is much more like "decimals going on forever without repeating": he found a way to make precise what that could possibly mean, using some previous ideas of another mathematician, Augustin-Louis Cauchy. As a result the construction is usually called "Cauchy real numbers," which is a little confusing but very respectful. Dedekind's idea was more like thinking about all the ways to slice a cake, rather than trying to find all the individual crumbs. If you find all the ways to slice it, you effectively have found all the individual crumbs, just somewhat indirectly. Both of these constructions give us a way to "fill in the gaps" in between the fractions, in a way that can be made entirely rigorous.

But my aim here isn't to explain Cantor's or Dedekind's construction of the "real numbers," which is what the number system including the irrational numbers is called. I just want to convey the idea that we have to understand things more deeply in order to teach them well, and that this type of understanding is one of the things that pushes the development of mathematics research. The work of Cantor and Dedekind making the real numbers precise is what enabled the rigorous development of the entire field of calculus, which in turn enabled all the developments of the modern world. And this all came from a couple of professors worrying about whether or not they'd be able to explain something to their students.

The importance of students' questions

To me, this all emphasizes the importance of students' questions. That's why I celebrate students asking searching questions, the questions that arise from them not simply accepting what we say, but wanting to know why, and where it came from: the questions that might sound innocent but turn out to be deep. Incidentally, this is very different from the kinds of posturing questions where students are just trying to prod or test their professors, or trying to catch them out or find their weaknesses. I'm afraid that students do that too (and probably more with women teachers, and most of all with non-white women teachers).

I don't blame them for doing it—a large part of the education system is set up to value and reward that sort of "smart," the sort where you "win" an argument by out-arguing someone, by forcing them into a corner, and leaving them unable to answer. Unfortunately this in turn creates an atmosphere in which teachers have to get themselves out of those corners, sometimes by telling people their questions are stupid. It's a vicious cycle that I wish we would stop.

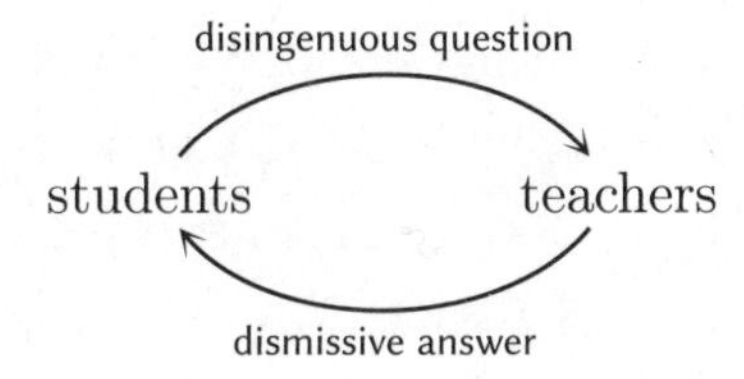

One way to stop it is to stop valuing that sort of zero-sum smartness (which depends on making someone else look stupid) and nurture the genuine, innocent questions instead. Too often we think there are "good" questions and "stupid" questions, but I prefer to classify questions into those where the person asking the question is genuinely trying to understand something, and those where they are trying to show how smart they are.

Questions where someone is genuinely trying to understand something are, in my opinion, always good questions, and often the most innocent ones are the ones that get me thinking the hardest. I love it when children wonder how many sides a circle has: Is it none (because there are no straight edges), one (all the way around), or infinitely many little tiny ones?

It might be tempting to think that these answers contradict each other and so one of them must be right and the others wrong. This is part of our unfortunate tendency to turn life into a zero-sum game, especially where arguments and disagreements are concerned, instead of trying to find the sense in which everyone is right, which brings us to deeper understanding. I would call the zero-sum approach "ingressive," and the seeking of deeper understanding together as "congressive," using the terminology I introduced in *x + y: A Mathematician's Manifesto for Rethinking Gender.* The approach seeking deeper understanding involves a lot more nuance than what is often associated with math.

Binary logic vs. nuance

There is a tension here. On the one hand there are many situations in which two apparently contradictory positions can each have a sense in which they're right, meaning that we are not in a binary right-or-wrong situation. However, mathematics is based on logic, and mostly it's based on binary logic, in which statements are either true or false. Am I now contradicting myself? I hope you can tell that I'm about to show a sense in which both of those things are true, that they're not actually contradicting each other. The point is that the nuance of having "senses in which" different things can be true operates at a different level from the binary logic we're using to build the foundations of mathematics.

The logic at the foundations of math is perhaps the fundamental place where I will concede that there are right and wrong answers in mathematics. But this is just about the basic logic we use to build arguments, it's not about the arguments themselves. Basic logic is built from logical implication, of the form "A implies B," where A and B are statements. This logical implication means "Whenever A is true, B is necessarily true." Another way of saying "A implies B" is "If A then B" (or, less succinctly, "If A is true then B is true").

There is a sense in which statements like this are binary but there is also a sense in which nuance is built in. They are strictly binary in the sense that an implication is either true or false. Either A forces B to be true, in which case the implication is true, or it doesn't, in which case the implication is false. Ambiguity can still be present, in that A might not be entirely forcing B to be true, but perhaps sort of nudging it. But in that case in mathematical logic the implication still counts as false. And that's the sense in which nuance is built in: because the nuance gets absorbed. For example consider the following statement:

"If you are human then you are a mammal."

This is absolutely true by definition, because humans are classified as mammals according to biology. However, now consider this statement:

"If you are white then you are rich."

Now, not all white people are rich, but some are. More to the point, white people are on average richer than Black people in both the UK and US and probably across the world. It might be tempting to say that the implication is sometimes true and sometimes not true, or true on average, but that's not how binary logic deals with it. The conclusion is only sometimes true, so according to basic logic the implication itself counts as false. That is what I mean by nuance being absorbed into the binary logic.

If the nuance is absorbed, you might wonder if it is lost. However, the important thing is that it hasn't been absorbed forever. Whenever we do abstraction in math we're not doing it permanently, we're just doing it as a temporary step to see what we can understand from the situation. We can then further refine our thoughts later. Even with binary logic we can keep exploring deeper and deeper into the nuance and express as much nuance as we want, for example:

"In the UK and the US the median income of white people is higher than the median income of Black people."

And then, as individual income is not the only indicator of wealth, we could look at household income, or household wealth. We could also look at access to resources such as education and health care. We could look at other indicators of inclusion in or exclusion from society such as voting, incarceration, and police violence. We could look at percentiles other than the median. We can keep homing in on as much nuance as we want.

That nuance is also built into the situation by context. That is, we might still have binary right-and-wrong answers in any given context, but there can be a lot of nuance in specifying what context we're in. It's like the fact that 1 + 1 doesn't just have one correct answer overall, but perhaps it does have one correct answer in any given context. Similarly "everyone is racist" might be true or false depending on what definition of "racist" you're taking, and considering this question focuses our attention on what we mean by "racist" rather than on the (probably less productive) goal of shaming individuals.

In summary, I concede that there is a sense in which math does have some definite concepts of right and wrong, because it's based on logic. Logic flows in particular directions, and going against those directions is not correct logic. However, this isn't the same type of right and wrong as saying that "1 + 1 = 2" and that's the only correct answer. The right and wrong of logic is more about right and wrong processes of deduction. For example, if we know that "A implies B" is true then whenever A is true we can correctly deduce that B is true. However, if we know that B is true that doesn't mean we can deduce that A is true, and if we do so then our logic is wrong. This happens far too often in arguments in daily life. For example, we know that everyone who is in the country illegally is an immigrant, as they must have come from elsewhere. But some people think this means that everyone who is an immigrant is in the country illegally. That is incorrect logic. There is no nuance about the right and wrong of that situation: the logic is simply wrong. On the other hand, someone might fear immigrants, they might object to immigrants, they might dislike immigrants. I would say that those points of view are ignorant, abhorrent, prejudiced, possibly bigoted, and often hypocritical, but they're not strictly logically wrong.

So I would say that math isn't really about getting the right answer, it's about building good justifications.

Justifications rather than right answers

College-level math tends to shift the focus from answers to justifications, which can be a shock to those who previously liked math because they found it easy to get the "right answers." In college the questions are likely to move away from "What is the answer to this question?" toward "Show that this is the correct answer." The "answer" is actually given in the question, so that there is no emphasis at all on what the answer is, only the justification.

We could shift the emphasis in this way for children as well. My favorite illustration of this is Christopher Danielson's excellent book *Which One Doesn't Belong?*. Each page has a set of four pictures, and the question "Which one doesn't belong?" But any of the four pictures could be the one that doesn't belong, depending on what sense of belonging you are picking. So there are no right and wrong answers, there are just senses in which different choices don't belong. This draws our attention away from the answer, and onto the justification instead.

I'd like to imagine doing this for something like times tables too. Instead of asking children "What is six times eight?" we could ask them "Show that 6 × 8 = 48." If we only ask "What is 6 × 8?" it is indeed possible that they just "know" this without thinking. I can intone "six eights are forty-eight" without really engaging any part of my conscious brain. But then if someone didn't believe me, I would be able to provide several different explanations to back up my answer, including these:

- I could count in eights: 8, 16, 24, 32, 40, 48.
- I could argue that $6 = 3 + 3$ so to find six eights I can do three eights and add it to three eights.
- Similarly $8 = 4 + 4$ so I could do six fours and add it to six fours.
- I could use the fact that six eights is the same as eight sixes, and for eight sixes I can use $8 = 10 - 2$, so I could do ten sixes and take away two sixes.
- Or $6 = 5 + 1$ so I could do five eights and then add eight.

This gets implemented in schools as children needing to learn different "strategies" for doing the same thing, and I often hear parents complaining how pointless this is because if they can do one way why do they need to know all these other ways (especially when the other ways are ways the parents themselves don't know about)?

The point here is that having different ways to think about something constitutes a deeper understanding of that thing, and it gives you more ways to check that what you're doing is secure. It's a bit like if you're building some scaffolding to climb up to the roof of your house. Before putting your life in its hands, you might want to check that the scaffolding is secure in various different ways, not just one way.

This is why it's important to see that math isn't just about getting the right answer, but about how you know it's the right answer. One problem with this is that times tables feature so early in typical math education, and people who are "good at math" are often quite fast at their basic times tables. This makes it look like they have memorized them, and then creates a contrived situation where it seems like memorizing times tables is a key part of being a good mathematician.

This is not the case at all. I have personally never memorized my times tables. My wonderful PhD supervisor, Martin Hyland, tells a story of his own childhood run-ins with times tables: Apparently when he was eight, his class was tested on times tables every day, and once any child got everything right three days in a row they could stop doing the tests. He was the only child in the class who never achieved that. He was also the only one who became a world-renowned research mathematician and professor at the University of Cambridge. As he puts it, he has a "poor memory for what seems meaningless" but a "good memory for the shape of ideas." Abstract math is about the shape of ideas, but unfortunately too many children see it as meaningless facts that you have to memorize.

I too am very bad at memorizing facts. I know my times tables, and can do them faster than average—but only up to 10. (Well, 11 I suppose.) But I haven't done it by memorizing them, at least, not in the sense of committing them to memory by rote. They are in my memory, in some sense, just like my name is also in my memory, but it would be quite bizarre to say that I memorized my name. I prefer to say that I know my times tables, or possibly even that I have "internalized" them, but really what I've done is understood various relationships between numbers so deeply that I can call them to mind very quickly using different methods, including mental visualization, and using the principles of commutativity (it doesn't matter what order we multiply), associativity (it doesn't matter how we group things together in multiplication), and distributivity of multiplication over addition. This is also what gives me more ways to explain things to people who don't understand, and I have always relished the chance to do that. This is why I am drawn to teaching, and this is why I am particularly drawn to teaching students who ask innocent questions, rather than students who think things are "obvious" and don't need explaining. Those supposedly obvious things often shed the most light on the process of doing mathematics, and if we gloss over them we miss so much of what I find profound and illuminating about math. One classic example is about dividing by zero. I'd like to end this chapter by discussing that, to bring together all the themes we've touched on about the framework of mathematics.

Why can't we divide by zero?

The question of why we can't divide by zero has been bothering people for generations. Some people think it's obvious: if we try and divide a packet of cookies but give each person 0 cookies, we will never use up all the cookies. But this depends on a particular interpretation of what "divide" means, and there is another interpretation: If we try and divide a packet of cookies between 0 people, how many cookies will each person get? This is a little tricky—it might seem like everyone gets 0 cookies. But also, everyone gets 1 cookie, that is, there are 0 people, and every single one of them gets 1 cookie. And also every single one of them gets 2 cookies. When we are considering "everyone" but there is a total of zero people in this "everyone" it's a situation called "vacuous": the conditions are vacuously satisfied because we're applying them to an empty set of people. It's like me saying that all the elephants in my house are purple: there are zero elephants in my house, and every single one of them is purple.

In order to understand why we can't divide by zero, we need a better understanding of what "division" is. Division is difficult. It is definitely harder than multiplication, which was already harder than addition. This is partly because there are two different interpretations of division as real-life scenarios.

For example, to do 12 divided by 6 you might take 12 playing cards and divide them out between 6 people. You might deal them out when playing a card game, so everyone gets one each, and then you start again at the beginning and give everyone another one each. Then at the end you see how many cards each person has, and you discover that the answer is 2.

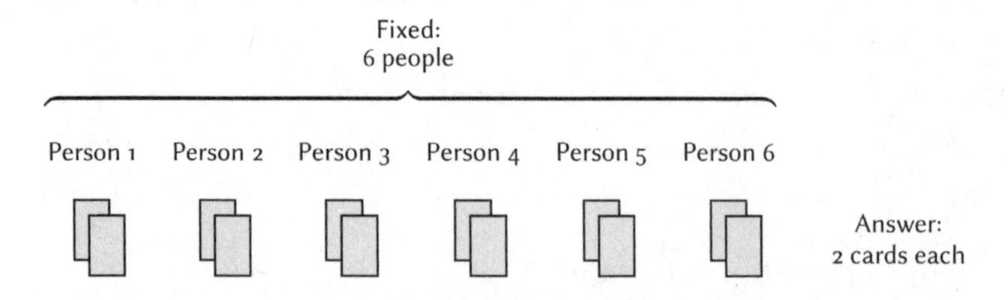

But there's another way of doing it: this time you take your 12 cards and you put them in piles of 6. So 6 of them go in one pile, and 6 of them go in another pile. Then at the end you see how many piles there are, and you discover that the answer is 2.

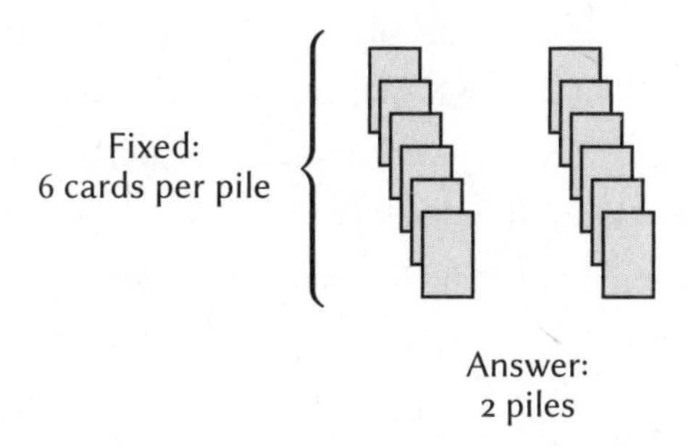

This difference can be very confusing. I was once helping a small child at school who was really struggling with division, and had a book to help her. The problem was that the book explained division by the first method above, and she really wanted to do it by the second method.

The answer is the same whichever way around you do it, but the processes are really quite different, as shown by the fact that in the first case the answer is "2 cards" and in the second it's "2 piles." In the first case you fix the number of piles and count how many cards are in each pile, and in the second case you fix the number of cards in each pile, and count the number of piles. It's not very obvious why those should give the same answer. At least I don't think it's very obvious, and it's mathematically not obvious, so I consider that the people who think it's not obvious are thinking more like mathematicians than the people who think it's obvious. And yet, the people who think it's obvious are more likely to be the ones who speed ahead with their homework and get praised, while the people who think it's not obvious sit down and think about it and appear to be slow at math.

Mathematicians don't define division in either of these ways, because there's too much ambiguity involved. We also don't define it in those ways because there's another way that we've already done, which means we don't have to come up with a new concept: we can do it like we did for negatives, where we thought about "canceling out," or rather, we thought about inverses. When we came up with negative numbers, we were thinking about inverses with respect to addition: we were canceling out the process of addition. For division we're going to think about inverses with respect to multiplication instead. Not only does this mean we can reuse some thought processes we've already developed, but it will also enable us to think more rigorously about the question of dividing by zero.

Division as an inverse

In rigorous mathematics, division is defined as the process of canceling out multiplication, just like we defined negatives by canceling out addition. This is saying that, abstractly, subtraction and division are the same thing. We have to be careful how we make this analogy though.

When we defined negatives we started by thinking of "canceling out addition and getting back to where we started," and this meant getting ourselves back to 0. We had already characterized 0 as the object that "does nothing" for addition, that is, if we add 0 to anything it doesn't change it. So we need to start by working out what is the "do nothing" object for multiplication. It's not 0 anymore, because if we multiply things by zero they do change—they all become 0.[†]

Instead, the "do nothing" number for multiplication is 1. That is, if we multiply anything by 1 it doesn't change. Again we can put this more formally by using the letter x to represent the number we start with. What we're saying is that for any number x, $x \times 1 = x$.

The technical term for this is identity. We say that 1 is the identity with respect to multiplication or the "multiplicative identity," and 0 is the identity with respect to addition or the additive identity.

† Yes, I know, 0 doesn't change if you multiply it by 0.

We can then ask about how we cancel numbers out to get back to the identity. For multiplication we might ask: How do we cancel out the number 4 by multiplication, to get back to 1? The number that does this for us is $\frac{1}{4}$. That is,

$$4 \times \frac{1}{4} = 1$$

This is how we *define* fractions abstractly, just like how we define negative numbers: we decide that we want multiplicative inverses to exist for all numbers in our world, so we include these as basic building blocks. "Dividing by 4" is then just a shorter way of saying "multiply by the multiplicative inverse of 4," just like "subtract 4" is a shorter way of saying "add the additive inverse of 4."

The thing is that this doesn't actually tell us how to find the answer to a division such as 12 divided by 6 in practice. It tells us in principle to do

$$12 \times \frac{1}{6}$$

which means "multiply 12 by the number that cancels out 6." In practice we typically then have to work out how to express 12 as "something times 6" so that we really can let the $\frac{1}{6}$ cancel out the 6. That is, if we can work out that 12 is 2 × 6 then we can do this:

$$\begin{aligned} 12 \div 6 &= 12 \times \tfrac{1}{6} && \text{by definition} \\ &= 2 \times 6 \times \tfrac{1}{6} && \text{re-expressing } 12 \text{ as } 2 \times 6 \\ &= 2 && \text{canceling out } 6 \text{ by } \tfrac{1}{6} \end{aligned}$$

If you think that seems very convoluted then I wholeheartedly agree, and that is the exact point I am trying to make. Division really is convoluted. Perhaps it now seems that doing it by sharing is much more straightforward, and that is also true, but it is limited. Defining division abstractly by inverses enables us to go much further with it, and extend it to other worlds such as shapes and symmetry where there really isn't a concept of sharing.

The abstract approach also enables us to understand things by analogy with negatives. For example, we can immediately understand the analogous result to $-(-x) = x$. What that said was that "the additive inverse of the additive inverse of x is x." But we could do that for multiplicative inverses as well. It goes in steps. First we declare that we're writing $\frac{1}{x}$ for the multiplicative inverse† of x, that is, the number that cancels out x back to 1 as in

$$x \times \frac{1}{x} = 1$$

Now what if we take the inverse of $\frac{1}{x}$? That will be this thing that strikes horror in so many math students' hearts: the question of dividing by a fraction.

$$\frac{1}{\frac{1}{x}}$$

But this is just "the number that cancels out $\frac{1}{x}$," and we already know that x does that. So we know that

$$\frac{1}{\frac{1}{x}} = x$$

This is "the same" result as $-(-x) = x$ at the following abstract level: the inverse of the inverse of x is always x, no matter what type of inverse you're doing (provided the inverse exists).

† As with additive inverses, we have to show that there's only one possible number that cancels out x, otherwise this definition would be ambiguous.

More generally, this explains why when we divide by a fraction we have to "turn it upside down," although it takes a few more steps to get to that from where we are so far. But what I really wanted to do was explain why this means "we can't divide by 0." Or rather, I want to explain in what sense we can't divide by 0.

Where we can and can't divide by zero

The question is: What would it mean to divide by 0? What we've just established is that division means "multiply by the multiplicative inverse." You might think that the multiplicative inverse of 0 is $\frac{1}{0}$, which doesn't exist. That's sort of true, but there's some logic to iron out: How do we know that $\frac{1}{0}$ doesn't exist? Can't we just throw it in as a building block, like we threw in $\frac{1}{2}, \frac{1}{3}, \frac{1}{4}$ and so on? We don't need to know what those are in advance, we just throw them in as building blocks and start playing around with them. This is a very good question.

The sticking point is that if we do that for $\frac{1}{0}$ we run into some problems. If this new object is going to be a multiplicative inverse for 0 then it's supposed to "cancel 0 out, back to 1." This means it's some number a such that

$$0 \times a = 1$$

But this can't happen, because $0 \times a$ is always 0. This means that 0 can't have a multiplicative inverse: there can't be a number that will satisfy the property in question, at least, not in the normal number system. And this is what it really means to say "You can't divide by 0": it means that in the normal number system, there is no multiplicative inverse for 0.

Another way of thinking about it is that the process of "multiplying by 0" can't be reversed, because it turns everything into 0. If we were going to undo that, we wouldn't know where to go back to, because everything has become the same. It would be like making a code in which you write every letter as X. Then I could send you this encoded message:

XXXX XX X XXXXXXXXXX XXXXXXXX

and you would have no hope of decoding it, because every letter has turned into the same thing.

Now, it's important to note that we've used another result here: we used the fact that $0 \times a$ is always 0. You might wonder where that comes from, which is a good thing to wonder, because that's rather profound as well. Perhaps if that isn't always true then we can divide by zero? That's an excellent thought. In fact, asking why we can't divide by 0 is, as usual, not the best question. A better question is "Where can't we divide by 0? Where can we divide by 0?"

We can't divide by 0 in the ordinary world of numbers, because the other rules of interaction would mean that dividing by 0 causes a contradiction. Where do those other rules of interaction come from? Well, they're part of the definition of the ordinary world of numbers. We'll come back to that, but for now I just want to stress that there are other perfectly valid mathematical worlds to explore, in which different things happen.

Mathematicians explore other worlds in which you can divide by 0, because, like many children, they get somewhat frustrated with the idea that you can't. We feel that there is a sense in which we can divide by 0 and get infinity, except that infinity isn't an ordinary number either. So we then need to be in a world containing infinity, and there are various ways of creating that. One of them is something I've done in my own research, where you basically just throw in infinity as a building block and take it from there. You just have to let go of some other rules of interaction, because once infinity is involved, those rules would cause some contradictions. For example, you might have to let go of commutativity of multiplication, the rule that says that multiplying things in different orders must give the same result. Or you might have to let go of some of the ways in which addition and multiplication interact. You might have to let go of multiplication being "repeated addition."

If this exploration is making you feel more and more confused then in a way you're doing it right. There is a lot of weird and confusing behavior that we uncover if we start really thinking about and questioning the things we've been told to take for granted. Making sense of that weird and confusing behavior is a central part of mathematics, a bit like when Europeans first encountered a platypus in Australia and were very baffled by this creature that appeared to be a contradiction. Of course, it wasn't a contradiction, it's just that the Europeans' worldview had been too narrow to allow for such a creature before. Getting our head around things that are confusing is an important part of developing our thinking and is one of the driving forces that motivates mathematicians, as we'll see in the next chapter.

CHAPTER 3

WHY WE DO MATH

Why isn't 1 a prime number?

One immediate answer is "because prime numbers are those that are only divisible by 1 and themselves, except 1 doesn't count." I hope you feel that this answer isn't very satisfying, because it's really saying "because the definition says so." That's another variation on the dreaded "Because I say so!" and provokes a follow-up question: Why does the definition say so?

This is something that perennially winds people up. It seems like such an annoying little caveat to have in a definition. It's the kind of thing students lose one point for on a test, and then they get irritated with the pedantry of math.

Maybe, inspired by what we've talked about already, you feel that you *want* 1 to count as a prime number, so you'd just like to include it and see what happens. That seems to be what we did with other things like 0, negative numbers, and so on.

But why might you want 1 to count as a prime number?

The question of 1 not being prime is a very good question. Because to answer it well, we have to ask ourselves why we're even thinking about prime numbers in the first place. What is the purpose of prime numbers? What is our motivation for studying prime numbers? What is our motivation for studying anything in math? In fact, what is our motivation for doing anything at all?

In this chapter I'll talk about why we do math, really. In school it can seem like we just do it to pass tests and get required qualifications. But mathematicians love math and care about it enough that they do it even when they have no more tests to pass. We do it because we have burning questions that haven't been answered yet. We do it because we really want to understand something more, because we don't want to accept someone else's answer on trust, or because we can see something shimmering in the distance and we want to get a better view of it. Sometimes it's because we have fit together some pieces of a puzzle and we're sure there are some more to fill in the gaps. Sometimes it's because there's a mysterious box and we just want to know what's in it. Sometimes it's because there's a mountain and we are drawn to climbing it to see the view. And yes, sometimes it's because we have a specific problem we want to solve. That's probably the most obvious motivation for math, but there's so much more to it than that. And sometimes we do it just because it's fun, because nurturing something that grows is fun, because finding sunlight is glorious.

Math is often pressed on us as a skill that we will need, but the math we present in school is often, in all honesty, not directly useful, so if you don't find it fun there really is no point doing it.

Pointless math

Periodically during tax season a meme goes around saying something like this:

> I sure am glad we studied triangles
> every time triangle season comes round

The implication here is that we slaved over triangles at school and it was completely useless, because we never need triangles in "real life," whereas we do need to understand taxes and so it would have been much more useful to study how to do taxes in school, rather than all that pointless stuff about triangles. (This is admittedly more widely relevant in the US where *everyone* has to do a tax return, unlike in other countries such as the UK where it's done automatically for people in regular employment.)

This meme makes me sad in many different ways at once. First of all, because it has an element of truth in it: that many things we do in math at school are not things that will ever be useful in daily life. Or rather, they won't be *directly* useful, and I suppose that is the real point. The point is that "useful" can mean a rather wide variety of things, and that we spend too long focusing on math as "directly useful" while simultaneously teaching math that is not directly useful.

There are two ways we could remedy that. One approach would be to teach math that is directly useful instead. I suppose this would mean things like taxes, mortgages, inflation, debt repayment, budgeting. Personally, I think that sounds awfully boring. It's also very limiting. Because if you teach How to Do Taxes, then it's not really applicable to anything except doing your taxes. Likewise there aren't many things that work quite like a mortgage, so understanding mortgages is not extremely helpful to anything except understanding mortgages.

Really this all comes back to why we do math education, which comes down to why we do math, as well as why we do education; and this comes down to why we do anything in life.

One of my favorite questions I've ever been asked at the end of a public math talk came from a six-year-old girl in Panama. She asked, "If math is everywhere, why do we have to go to school to learn it?" This question encapsulated what I find wonderful about innocent questions, at both a mathematical level and also at a meta-level of language. My Spanish is exceedingly rusty, but I was able to understand her question as posed in Spanish; however, I had no chance whatsoever of being able to provide an answer in Spanish, so had to rely on the interpreter for my response.

Innocent questions in math are like that: they can be very easy to pose and very easy to understand, but extremely difficult to answer.

To me, the point of formal education, as opposed to life education, is to accumulate knowledge from generations and generations of humans without having to go through the entire process ourselves to learn it "from experience." Yes, some things can only really be learned from experience, like, perhaps, how to deal with grief. But even in that I have been helped immeasurably by an expert psychologist and all the formal knowledge she brought from the field; however, one part that you can only learn by experience is how you as an individual are going to respond to the pain and also to the interventions.

We can get a glimpse of so much more during formal education than we could if we just waited for the experiences to happen to us. This does provoke the question of why (or whether) that's a good thing, and I'll come back to that in the next chapter.

So personally I believe it's most powerful for formal education to address things that are not too closely related to real life, but that are very broadly transferable instead. General foundational skills, if you like, rather than very specific ones.

That's a very brief account of why we do education; what about why we do math? Why do we do anything?

People do things because they're useful, or because they're fun, or possibly because there will be some dire consequences if we don't do them. (I realize that this does not include darker motivations like revenge, anger, hatred.)

Perhaps fun *is* useful? This comes back to my earlier point about different meanings of the word "useful." There's the rather utilitarian version of direct usefulness, but then there's the other version that is more transferable. So rather than "I am doing this thing that I can then do to great effect in my life," it's more like "I am doing this thing which is exercising my brain in a particular way so that I can then use my brain to great effect in my life."

So the question then isn't "Will I ever use this exact thing in my life?" but rather "In doing this, am I developing myself in some way that will be beneficial later?" I find that this latter definition of "useful" is more . . . useful. It's also more relevant to why we do math. Thus, if we're doing algebra or thinking about triangles or prime numbers, the point isn't that we will need algebra or triangles in our future daily life; the point is that we're developing our thinking in a way that will enable us to think more clearly about daily life in the future.

Now, there are instances where the exact thing we're studying will be useful in our future lives. There was another meme that went around during the COVID-19 pandemic, depicting a math teacher teaching a class about exponentials, and some bored students saying "When are we ever going to need this in life?" Unfortunately, when the pandemic began it would have been really helpful if more people had understood exponentials in advance. Instead, when scientists tried to point out that it looked like things were going to get really bad, because of how exponentials work, far too many people thought they were scare-mongering or making things up when you "can't predict the future."

So I'm not trying to say that school math is never or should never be directly useful. And later in this chapter we'll see that some things that mathematicians did mostly for fun actually turned out to be very directly useful later—it turns out that humans are not very good at predicting what will be useful in the future.

In this chapter we're going to look at various different motivations for the math we do. This isn't just about why we do math, it's about why we do it in the way that we do it. There are some deep guiding principles at work, stemming from our view of math as "the logical study of how logical things work." An important part of studying things logically is to take it very slowly, and understand what the basic building blocks of the situation are, and how they interact with each other. We've already seen some of this in the first two chapters. We'll also see that understanding the principles at work doesn't just help us get "the right answer" (although it might also do that); it helps us understand more situations at once, and helps us move on to understanding much more complicated situations analogously, by a process of mathematical generalization.

To address the question of why 1 isn't prime, we need to think harder about the principles of prime numbers rather than just their definition. And the principle is that they are basic building blocks for numbers.

Seeking basic building blocks

The deeper reason that mathematicians are interested in prime numbers is because we're interested in building blocks. We said in the previous chapter that we're interested in breaking down large ideas into small ones and seeing how they can all be built up from small ideas, or small building blocks.

We've been thinking about the so-called natural numbers 1, 2, 3, . . . and we've talked about the fact that we can build them "freely" starting with just one basic building block: we just start with the number 1 and use addition as our building method. The resulting structure is quite simple as we only need one building block and we can build the whole thing. I don't mean that numbers themselves are simple, I mean that from the point of view of this building process they are straightforward.

We could give ourselves more building blocks, but that would be redundant. Perhaps we want 2 to be a building block as well—but we can build 2 as 1 + 1, so we don't really *need* to start with 2. It's like wanting to travel light and take as few things with you as possible. I'm not a dogmatically light traveler when it comes to physical traveling, as I personally want to balance lightness with comfort and fun. But in math I do enjoy the principle of investigating how light we could travel if we really wanted to. That's the principle of finding basic building blocks. We simultaneously want enough blocks to be able to build everything in our world, while also having nothing redundant.

Those two aims work in opposition to each other: if we take more things, we're more likely to be able to build everything, but also more likely to have redundancy. If we take fewer things, we're less likely to have redundancy, but also less likely to be able to build everything. So we need to find a balance with not too much but also not too little.

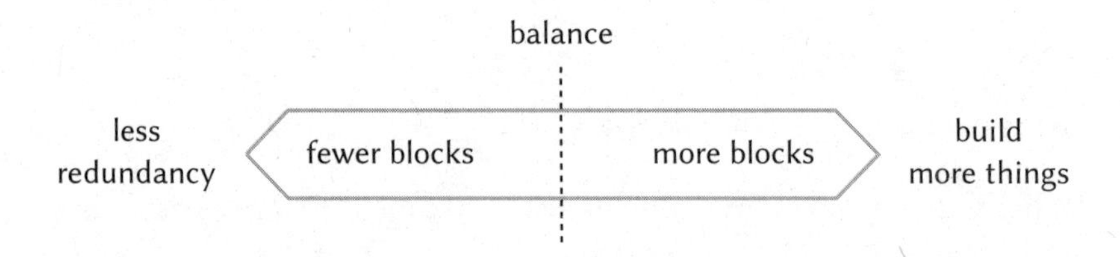

In the extreme, we could take *everything* to be a basic building block, and we'd be sure of being able to build everything. But there would be a whole load of redundant blocks that we could build from other things. At the other extreme we could take *nothing* as a basic building block, in which case nothing would be redundant, but we also wouldn't be able to build anything.

In terms of personal beliefs, these are actually two extreme ways of being completely rational. In the first case, you take every belief of yours as a fundamental belief. But you haven't really achieved anything because you haven't put in any effort to see how any of your beliefs follow from other ones. In *The Art of Logic* I called that logical but not powerfully logical. It's not at all usefully logical.

At the other extreme some people are determined that they take *nothing* as a fundamental belief, because they think being perfectly rational precludes taking anything on faith rather than deducing it logically. However, if you do that you won't technically be able to deduce anything, because you can't deduce something from nothing. If you go around insisting that you don't think anything is true then you might well be perfectly consistently rational, but again I don't think that would be particularly helpful.

What we try and do in math is increase the number of building blocks just enough so that we can build everything, but not so much that we include redundant ones. Or, we could look at it in reverse—we might discard beliefs that are redundant, and keep discarding them until we only have the ones that we really can't build up from other ones. Those two processes will, in theory, meet in the middle, at a perfect balance of being able to build everything but not having any redundant blocks. Finding that balancing point is very satisfying, and is a key part of understanding any mathematical structure deeply, and deep understanding is what we're always seeking in math.

This is what we're trying to find with prime numbers.

Prime numbers as basic building blocks

The world we're investigating is still the world of natural numbers 1, 2, 3, . . . , but we're now looking at them in a different light. Previously we investigated building them by addition, but now we're building them by multiplication. This is more complicated than building them by addition, because, deep down, they are *defined* by addition. So we're taking something that's organically grown out of addition, and seeing if we can fabricate it ourselves from multiplication. It's a bit like the chemical synthesis of organic substances, or the electronic imitation of musical instruments on a synthesizer.

We can also view this as an investigation of the interaction between addition and multiplication. It's a bit like wondering how two different friends of yours will get on, or wondering if two different animals are biologically capable of mating (and what happens if they do).

This brings us to what prime numbers are really: they're going to be the basic building blocks we need if we're building by multiplication. My preferred way to think about them is that we come up with this idea first, and *then* we work out which numbers qualify. This consists of working out which numbers we really need, and which numbers are redundant. That's where the usual "definition" of prime number comes from. I prefer to call it a characterization, because really we're characterizing which numbers count as good basic building blocks in this particular context.

And here's the thing: 1 is not a good building block. Actually 1 is completely useless as a building block by multiplication, because if we multiply things by 1 nothing happens. It's a non-building block. Indeed, that's how we defined it: it's the "multiplicative identity," meaning that multiplying by it does nothing at all. Identities are therefore always excluded from being building blocks because they're useless for building. (This doesn't mean they're useless overall—in fact they're very important, just not for building things. I like to think that I'm useful to society although not when it comes to building houses.)

That really deals with why 1 isn't a prime number, but now that we're here we might as well finish this off and characterize the things that are prime numbers. The point is that all the bigger numbers do help with building, but some of them are redundant: any number that can be made by multiplying two smaller numbers together is redundant as a building block, because we can always build it from the smaller numbers. So we don't need 4 as a building block because we can build it as 2×2. We don't need 6 because we can build it as 2×3. And so on.

In summary, the productive building blocks are all the numbers except 1, and the redundant ones are all those that can be expressed as a product of two smaller numbers. Thus we are left with "all the numbers with no factors except 1 and themselves, and 1 doesn't count."

There is one final technicality: you might wonder how we're going to get 1 in our world at all if we don't include it as a building block. Mathematicians do this by considering "do nothing" as a valid building *process*. This is how we include 0 even when we're building by addition: because we are allowed to "do nothing" and that leaves us at 0. When we're building by *multiplication*, "do nothing" leaves us at 1.

So we get 1 in our world not by including it as a building block, but as a process.

Here is a slightly more technical explanation in case you're interested (we won't need it again so feel free to skim it and not worry if you find it hard). Another way of thinking of our building blocks is that if we had a huge long list of all the prime numbers, we could make a recipe book for every natural number, by going down the list and saying how many copies of each prime number to throw into the mix. So for 6 we'd say "throw in one 2 and one 3 but nothing else." For 10 we'd say "throw in one 2, no 3s, one 5, and nothing else," because $10 = 2 \times 5$. For 8 we'd say "throw in three 2s" and remember that we're building by multiplication so throwing in three 2s doesn't mean we're doing $2 + 2 + 2$, it means we're doing $2 \times 2 \times 2$, which is indeed 8.

Here's a table of some entries in this recipe book. The "ingredients" (the prime numbers) are listed down the side. The number we're making is at the top, and then each column of entries tells us how many of each prime number to throw into the mix.

ingredients	how to make ... 2	3	4	5	6	7	8	9	10
2	1	0	2	0	1	0	3	0	1
3	0	1	0	0	1	0	0	2	0
5	0	0	0	1	0	0	0	0	1
7	0	0	0	0	0	1	0	0	0

Remember that repeated multiplication is written as an exponent, as in 2^3. So when we say "throw in no 3s" that means we're doing 3^0, which is 1. So the recipe for 2 listed in the table is telling us:

$$2 = 2^1 \times 3^0 \times 5^0 \times 7^0$$

I actually keep recipes in a big spreadsheet like this, with ingredients listed down the side. It enables me to scale recipes up or down without having to do mental arithmetic, and also to compare different recipes for the same thing. It also means I can calculate a shopping list if I'm having a party with five or six (or more) desserts.

Now the question is: can we make a recipe for 1 with these ingredients? Well, we can just throw in nothing at all. In terms of the table, this means we'd have a whole column of 0s. So we're doing 2^0 and 3^0 and 5^0 and everything is to the power of zero. If we multiply all those together we do indeed get 1, so we don't need to include it as a building block; it can go in the table as the first column like this:

		how to make …									
		1	2	3	4	5	6	7	8	9	10
ingredients	**2**	0	1	0	2	0	1	0	3	0	1
	3	0	0	1	0	0	1	0	0	2	0
	5	0	0	0	0	1	0	0	0	0	1
	7	0	0	0	0	0	0	1	0	0	0

This an example where going up to a higher level of abstraction makes things make more sense, provided you've accepted the higher level of abstraction. And it's a good level, because it turns out that a lot of other "null" situations work similarly, where we need to think about doing something "zero times" and get confused about what that could possibly mean. I'm not just trying to understand prime numbers and the number 1, which is admittedly a rather specific scenario. Rather, I'm always seeking to understand things as deeply as possible and in a way that will help me with other things at the same time.

There is a little more to be said about these "recipes," which is that for each natural number there is only one possible recipe for building it from our basic building blocks. This is the point of eliminating redundancy—it removes ambiguity from our recipes. This idea is encapsulated in what is called the "Fundamental Theorem of Arithmetic," which says that every natural number can be expressed as a product of prime numbers in a unique way.

The fact that every natural number *can* be expressed as a product of primes is what tells us that we have enough building blocks to build everything. The fact that the expression is unique is what tells us we don't have any redundant building blocks. We have to understand "unique" to mean that changing the order of the factors doesn't actually make a different recipe. For example we can express 6 as 2×3 or 3×2, but those count as the same recipe. (They give the same entry in the table of recipes.)

Now, note that if we included 1 as a building block we'd also be able to say $6 = 3 \times 2 \times 1$ or $6 = 3 \times 2 \times 1 \times 1 \times 1$ or include any number of 1s at all, so our recipe wouldn't be unique anymore. This could be thought of as a reason for not including 1 as a prime number, but I prefer to think of it as a reason for eliminating redundancy, which in turn is our reason for not including 1 as prime.

One of the things that can make math confusing is that there are usually many different ways of telling the same story, and different ways resonate with different people. One way to tell this story is to define prime numbers and then prove that the Fundamental Theorem is true. But I prefer to tell the story a different way: I think of the Fundamental Theorem as being the goal, and the definition of prime numbers as being what we have to do to make the theorem true. To me math is more like having a dream of something you want to be true, and then working out what you would need to do to make that dream true. There might be various different ways of making that dream true, and you can study them all separately to see what they're like.

Incidentally this is how I operate most of my life as well, by dreaming up what I would like to be true about my life and the world at large, and then working out what I would have to do to make that dream come true. Often what I would have to do turns out to be rather unrealistic (or even impossible) but thinking about the dream in that way at least helps me understand something about it, and then maybe helps me understand how to make some part of it true, or how to make it at least a bit *closer* to reality than what I have now.

Abstract math is often motivated by dreams and desires. We dream of something we want to do, we dream up a world in which we can do it, we investigate what will make it happen, and we make definitions and building blocks to create those dream worlds. This is quite different from how math is usually presented in mainstream education, and the problem is that we tend to mix up the different purposes of math education.

What is math education for?

Broadly speaking, I believe in three reasons math is an important part of education. First there's the possibility of direct usefulness. Then there's the fact that it is crucial as a basis for further study in various subjects, including higher math, but also most of science, engineering, medicine, economics. Many people are not going to go into any of those fields, but we shouldn't close those fields off to anyone too soon.

The third reason for math being part of education is the *indirect* usefulness, the fact that it's a way of thinking that can be powerful in a very transferable way. This is the aspect that is the most broadly relevant, that is, relevant to the most people—because it's relevant to everyone. But it's also somehow the least emphasized. That's the wrong way around. If we emphasize this aspect rather than "direct usefulness" then it becomes much more possible to see why we study anything, including triangles, and it will help shift our focus to those deeper aspects of why we do it, rather than the more limited aspect of "direct applications."

This is something like doing exercises to strengthen our core muscles. There is no activity in life that involves *only* core muscles, but it's helpful to have a strong core because it enables us to use the rest of our muscles to greater effect. It gives us better access to the rest of our strength, as well as protecting us against things like balance issues, tripping over things, and hurting our back. One key here is that a strong core means we can better engage our other muscles, making us stronger without having to enlarge all those other muscles directly.

Arguably, the most useful part of math is the part that operates like core strength for our brain. It's not that what we learn is specifically directly applicable to anything, it's that we have strengthened our brain in a way that makes us better able to use the rest of our brain, without having to directly train those other parts of the brain.

For example, thinking about different types of sameness for triangles has helped me get good at being able to do abstraction and see connections between disparate things, and that abstract skill is very widely applicable, even if the stuff about triangles isn't. Building up arcane-sounding arguments about how triangles fit together has helped me practice the general skill of building up logical arguments, and that general skill is extremely useful.

On the other hand, there are parts of math that *used to be* directly useful but that really aren't anymore. If they're also not indirectly useful, I don't believe there's much purpose to learning them these days. For example, my parents' generation were still learning to use a slide rule at school. That's an old device from before calculators, which used the theory of logarithms to make a handy little tool for multiplying large numbers without needing a proportionally large device. I don't think anyone argues that it's still important for us all to know how to use one of these things. It's just like the fact that at some point in the past horseback riding was an important skill, but now that we have cars, learning to drive a car is much more important. Even driving a car isn't exactly crucial, but riding a horse is even less crucial as a general life skill. Of course, it's very important to certain professions, and it's fun to many people, but it's not extremely transferable, although I'm sure it does teach some good life skills.

However, not all "old-fashioned" school math is quite so obsolete. Adding numbers in columns has definitely gone out of fashion, but I see slightly more of an indirect point to it than the slide rule.

Why learn column addition?

Adding in columns comes into play when we're trying to add numbers larger than 10, say maybe 153 + 39. Depending on your era (and country) of math education, it might seem very natural to you to line them up in columns and then start adding from the right:

- 3 + 9 is 12, and now we have to carry 1 into the next column.
- Moving one column left, we have 5 + 3 and we also have the carried 1, which makes 9.
- Finally in the first column we just have 1.

$$\begin{array}{r} {\scriptstyle 1} \\ 1\,5\,3 \\ +\,3\,9 \\ \hline \mathbf{1\,9\,2} \end{array}$$

This is very much an algorithm for adding numbers, and you might think it's not very useful anymore, now that we all have calculators with us all the time (on our phones, for example). Personally, I use the calculator on my phone a lot, because although I am perfectly fine at mental arithmetic, I don't enjoy it, nor can I be sure I'll always get it right the first time, plus it tires me out cognitively and requires me to switch my brain function. For example, when splitting a restaurant bill after dinner with friends, my brain will be fully in social mode, and I won't want to take it into arithmetic mode in order to do a boring calculation. All that is to say that sometimes I'll add numbers in my head, but sometimes I'll just cut to the chase and use my calculator.

One argument is that we should still know how to add things up without a calculator just in case we get stuck without one. That's a rather tenuous argument, like saying we should all know how to ride a horse just in case we get stuck somewhere without a car (but with a horse).

I did get attacked on social media once for this, with someone saying I was not taking into account the possibility of someone not having enough money to charge their phone, so having no access to a calculator, but needing to buy groceries with their last remaining cash, so needing to be able to add up their groceries and not wanting to wait for it to be done at the register as it's humiliating to have to put things back because you don't have enough money.

I agree that this would be a tragic situation, but I would hesitate to base any math education on a preparation for that eventuality. I would rather have math education try to help people have more chance of avoiding such situations, as well as working on society's structures so that nobody ever is forced into such situations (plus also providing free phone charging stations and self-service checkouts).

Anyway a big criticism of this algorithm for adding in columns is that, like most algorithms, it allows someone to bypass a large part of the understanding of the situation. This is, as I believe people say, a feature, not a bug. That is to say, this shouldn't just be considered as a *bad* thing about algorithms. It's the whole point of algorithms: to enable us to do something more or less on autopilot, and save our brainpower for things that are more nuanced. "Autopilot" is usually taken to be a bad thing when we use it as a metaphor, but we should really appreciate autopilot for what it is. After all, it's fantastic that pilots can let the autopilot take control so that they can rest, and then when they need to take over because something subtle is happening or there is an emergency, they can be much more fresh and alert, having not had to do all the very routine stuff earlier on.

Where column addition is concerned, educators now generally understand that there are ways of adding multi-digit numbers that engage (and thus nurture) a more meaningful understanding of what is going on with those numbers. Children are now typically taught many different "strategies" for doing this, sometimes to the bafflement of the adults in their family who would rather just add these numbers in columns and be done with it. For example, for 153 + 39 a child might be encouraged to notice that 39 is almost 40, and it's probably easier to do 153 + 40 in your head than in columns. Then once you've done it, you then say "oh, but that was adding on one too many" and you take one off again.

Another way would be to go in exactly the opposite direction of column addition: start with the 100, then notice that we've got 50 in one number and 30 in another, so that gets us to 180. Now what remains is 9 and 3, which is 12, so we add that on to our previous total to make 192.

It's true that all of this does encourage a deeper understanding of the interaction between numbers, but it has to be done rather carefully, because if children are still focused on getting the right answer, they will be extremely frustrated by getting the right answer one time and then having to do it again and again.

I believe in the value of seeing all those methods, but I also think that column addition has a deeper point. It is more related to the encoding of numbers in columns in the first place, which is really a rather wonderful idea. Imagine if we had to come up with a completely different symbol to represent every number. We would have to think up infinitely many different symbols for the numbers, which would be impossible. Instead, people came up with a very handy way that we could represent *all* numbers, using only 10 symbols. This began with "counting rods" in China over two thousand years ago, and was followed by the Hindu-Arabic system that we're now familiar with, using the digits 0, 1, 2, 3, 4, 5, 6, 7, 8, 9. (Earlier cultures used systems with other numbers of digits such as 16.)

The system is really like using an abacus where you use the first row of beads to count up to 10, but once you get to 10 you slide one of the next row of beads along to remember that 10, and start the top row back at 0. If you count another 10, you slide another of the second row of beads to remember two sets of 10, and start the top row back at 0.

Here is a picture of my childhood abacus, which I have treasured for all these years. If the top row is the 1s, the second row is the 10s, and the third row is the 100s, this configuration is representing 231.

This idea is ingenious, and I don't think we draw enough attention to how ingenious it is. The Romans did not use this system for representing numbers, and as a result Roman numerals are much more convoluted, with placement sometimes representing addition, as in XI, and sometimes representing subtraction, as in IX. The Romans were very developed in many ways but they are not known for great advances in mathematics.

Thinking about the ingenious use of digits might lead to questions like: Do we really even need 10 symbols? Could we do it with 9? How many do we really need? This then relates to the question of finding minimal building blocks, and also to the question of generalizing to a broader context.

Generalization

In life, "generalization" can mean that you're making a sweeping statement, or making an assumption about a group of people based on only a small number of representatives. That is often counterproductive or insulting, or even dangerous. But in math "generalization" means carefully expanding your context so that you can include more things in your understanding. When we generalize in math, we do start with a smaller world and expand to a bigger world, but we don't assume that the bigger world will work the same as the smaller world. Instead, we try find a way to express something that's going on in the smaller world in such a way that it really does hold in the bigger world as well.

For our use of 10 digits, that means instead of saying something like "all multiples of 10 end in 0," we consider the possibility of an abacus with a different number of beads on each rod. Then we might notice that if we had 9 beads per rod, then all multiples of 9 would end in 0, and with 11 beads all multiples of 11 end in 0. Then the generalization is that with n beads on each rod, all multiples of n end in 0. Again, we've used a letter, n, to represent an unspecified number so that we can theorize about all numbers at once.

Why might we want to generalize this situation at all? There's no particular reason to choose 10 as the base number of digits, but the obvious "explanation" for it is that we have ten fingers, and these are the most natural objects we carry around with us to help with counting. However, this is more of an emotional explanation than a historical or logical one (but it also "explains" the fact that we use the same word "digit" for both the symbols and our fingers). Some cultures used different base numbers of digits, such as the Mayans who used 20 (perhaps including their toes?). Some traces of base 20 can be seen in French when we get to 70 and above, where the numbers translate literally as "sixty-ten, sixty-eleven, sixty-twelve..." up to "sixty-nineteen," followed by "four twenties" instead of "eight tens" for 80, and so on.

Some Native American cultures base their counting on the number 8, in some cases because of using knuckles, for example in the Northern Pame language in Mexico, or the spaces between the fingers, such as in Yuki, a Native American language in California. There is an argument for using 5 instead, and using both hands like a two-row abacus, so that we could count up to 25 on our fingers rather than just 10. That's an argument from the point of view of physical logistics, but from the point of view of abstract logistics, 60 is a really excellent starting point on which to base a number system, as it has so many factors. Whereas 10 only has 2 and 5 (other than 1 and itself), 60 has 2, 3, 5, 6, 10, 12, 15, 20, and 30. This was used to great effect in Babylon several thousand years ago. This, together with the Babylonians' use of place values for representing numbers, arguably contributed to Babylonians being very advanced in mathematics compared with the ancient Egyptians or the Romans.

I've been using the word "base" here informally, but it is also technical: when we pick a base number of digits to use to represent our numbers, that really is called a base in math. So our usual way of writing math with ten symbols is called "base 10" whereas if we only used nine it would be called "base 9." We can use any whole number of symbols bigger than 1 (because, again, 1 wouldn't get us anywhere). So we could most minimally do this with only two symbols, and the resulting system is called binary. By the way, when we use ten digits, we usually write them as 0, 1, 2, 3, 4, 5, 6, 7, 8, 9. So when we're using just two digits, we usually write them as 0, 1. As a result, binary numbers end up looking like strings of 0s and 1s.

This exploration of bases is both indirectly useful (in how it gets us thinking) and also directly useful. We can use our fingers as binary digits, by using the two positions "up" and "down" to represent 1 and 0, enabling us to count all the way to 1023 on our fingers, at least in principle (though it's rather tricky in practice). I have also used birthday candles as binary digits, with "lit" and "unlit" representing 1 and 0; this means that with just seven candles I can celebrate any birthday up to 127, which should do for a while as the oldest verified living person is currently "only" 118.

Unfortunately it does take rather a lot of concentration so it might not help much if you're trying to use your fingers to keep track of a number without engaging too much of your active brain. However, that's not an issue computers have to worry about.

Binary is put to very powerful use in computers, a somewhat less esoteric use than the ones above. For computers, rather than using fingers or candles, we implement the amazing idea that we can use electrical switches as binary digits, with on and off representing the digits 1 and 0. Building computers from tons of binary switches is a rather powerful direct use of this apparently arcane-sounding thing about how we represent numbers.

In a way, that's a slightly weak link with the idea of adding in columns, because we can learn about the columns (usually called place value) without having to add up in columns. But adding up in columns emphasizes the process of adding up the parts that correspond to each other. We have to be careful to line the matching columns up, and not do it like this by mistake:

$$\begin{array}{r} 1\ 5\ 3 \\ +\ 3\ 9 \\ \hline \end{array}$$

This leads to the general idea of gathering like things together to add them up. For example if we have one bag with 2 bananas and 3 apples, and another bag with 5 bananas and 1 apple, and we're putting them together to see how much fruit we have, it's quite natural to group the bananas and apples and say that we have 7 bananas and 4 apples. It would be more unusual (perhaps indicative of a different type of thinking, or a resistance to commutativity) to say that we have in total 2 bananas and 3 apples and 5 bananas and 1 apple.

Lining up the numbers in columns engages us in the discipline of recognizing that the columns represent different things. It means we sort of have to think of 153 as not just a string of symbols but 1 of something, 5 of something else, and 3 of something else. In fact it's 100s, 10s, and 1s, so the columns are really saying this:

100s	**10s**	**1s**
1	5	3
	3	9

Similarly in the situation with the fruit we could make a table like this:

bananas	apples
2	3
5	1

and then add up the columns. The subtlety with the numbers is that if we get enough 1s together, they spontaneously turn into a 10. This doesn't happen with the bananas and apples—there is no way to take a quantity of apples and spontaneously turn them into a banana. That is, there isn't usually a way to do that, but I do remember a fairground when I was little, with games where you could win different levels of prize; if you won enough prizes at one level you could then trade them in for a better prize. This would be like getting together enough 1s to turn them into an entry in the next column.

The scheme with columns has combined two different principles: the abacus principle, where a certain number of things at one level counts as one thing at the next level, and the "gathering like things together" principle. The latter is then very important in algebra where rather than apples and bananas we might have x's and y's and we group them together. We might then be adding expressions like $(x^2 + 3x + 1)$ and $(2x + 4)$. Instead of 1s, 10s, and 100s, these expressions contain 1s, x's and x^2's. We can then add them in columns in the same way as before, taking care to note that it's more like apples and bananas—no matter how many x's you get in a column, they never turn into an x^2:

$$\begin{array}{r} x^2 + 3x + 1 \\ 2x + 4 \\ \hline x^2 + 5x + 5 \end{array}$$

The context is new but the principle will be familiar if we've done some column addition.

The moral of this story is that sometimes an algorithm leads us into some interesting thought processes, even if the algorithm is a bit of an anachronism that we don't really "need" in math these days. A more contentious example of that is long division, which is an algorithm for dividing large numbers by numbers with more than one digit. What tips long division over the edge into "not worth it" for me is that it isn't even a very good algorithm, and in the next chapter I'll come back to what makes for "good" math at all. It's true that long division is *slightly* transferable to other parts of math (such as long division in algebra) but I consider that to be a somewhat tenuous piece of transferability, especially when weighed up against the fact that the algorithm isn't very good, causes a lot of students a great deal of pain, and that it doesn't really illuminate very much about the situation. That's why my vote is for ditching long division, but keeping column addition *provided* it's mostly a vehicle for further discussion about how numbers work, not an important method for getting the right answer, and certainly not a rod to metaphorically beat children with if they can't perform the algorithm smoothly.

The upshot is that some things in math are just useful techniques but others give us actual insight into what is going on. (Mnemonics fall very heavily in the first category with no insight at all, and we'll come back to some of those later.)

But there's another reason we do math, which is just sheer curiosity and possibly fun, like children jumping in puddles.

Jumping in puddles and climbing mountains

Children often love jumping in a puddle if it's there. I admit I also like doing that as long as I know that my shoes are very waterproof and nobody's close enough to get splashed. Since living in Chicago I've bought my first pair of wellies since childhood (I'm not even sure what they're called in the US). They're actually not for when it rains, although that can cause very large puddles with the bad drainage here.

It's for when the snow melts, which causes ankle deep riverlets along the pavement, and sometimes practically a lake pooling around where the crossing is, exactly where we hapless pedestrians need to step to get across. That's when I really need my wellies.

And I admit that when I am wearing them it is really fun to splash around in those little ponds, and stomp in puddles. As I said earlier, in many ways I'm just a small child who never grew up. When it has just snowed, I will put my snow boots on and take detours walking on the side parts of the pavement where nobody has walked yet, so that I get to walk in fresh snow, just for the sheer joy of it.

Sometimes math is like jumping in puddles and taking joy in fresh snow. And sometimes it's about climbing a mountain, just because it's there, just to see if we can.

I have never been drawn to climbing mountains myself (I'm very averse to physical danger) but perhaps I understand the urge to do it, because sometimes the urge to do math seems like that. It's just that the mountains are abstract. But we're still drawn by curiosity, to see what's there, and to see if we can do it. Some mathematicians are drawn by something more like a conquering urge, that there's an unanswered problem and they simply want to solve it. My motivation is less about conquering and more about wanting to shine light, clear away fog, and see more clearly, like climbing a mountain to admire the view from the top.

Sometimes we do math out of sheer curiosity, and pursuing something out of sheer curiosity can be really fun (as well as irresistible for some people). And sometimes it's just fun because it feels satisfying to slide things into place, like a jigsaw puzzle. I know some people don't like jigsaw puzzles, but it does seem that plenty of people enjoy them even though they wouldn't say they enjoy math.

There's an xkcd cartoon I really like that shows someone pulling a lever on a mysterious machine and getting a painful electric shock as a result. The cartoon then branches off like a flowchart. On one side it shows "normal people" with a thought bubble saying "I guess

I shouldn't do that." On the other side it shows "scientists," with the thought bubble on that side saying "I wonder if that happens every time."† The scientific urge is to keep testing things to understand their behavior.

Mathematicians are like that with explanations. If something is unexplained, or an explanation is unsatisfactory, we just want to keep prodding it, or digging around in it, or exploring it, to see what's really going on. If I'm trying to go somewhere and I get lost, then I will always study a map later to work out what happened. I only recently realized that not everyone feels this urge. I have an endless impulse to understand things further.

Sometimes this takes the form of getting to "first principles," like wanting to bake things from really basic ingredients.

First principles

I love making tiramisu, from eggs, sugar, mascarpone, coffee, brandy, savoiardi biscuits. At some point I decided I wanted to make my own savoiardi biscuits rather than using bought ones. Then I wanted to make my own mascarpone. I have not gone as far as raising chickens for fresh eggs, milking cows, or making my own brandy. We all have different concepts of what counts as "first principles," both in the kitchen and in math.

The beginning of an undergraduate math course is often quite a shock to the new students. Those who choose to do math in college are typically people who've done really well in math at school, were probably the best in their school at math, and have always found it easy. To those people, the first things we do in college-level math can seem simultaneously trivial and arduous. We set about proving some rather basic things from first principles, things that the successful

† The title is "The Difference" and it can be found at https://xkcd.com/242/.

math students have considered to be "obviously true" for years. For example, we prove that if you multiply any number by 0 you get 0.

This is something I've seen many children struggle with in the early years of elementary school. It seems to separate children out—some find it really obvious, and others don't understand why it's true. This feeds into the erroneous idea that there are "math people" and "not math people." But the thing is that mathematicians think it's *not* obvious why multiplying by 0 gives 0, which is why we feel the urge to prove it from first principles.

It all comes back to our desire to understand the entire number system with as few building blocks as possible. You might think that multiplication is "repeated addition" and so multiplication by 0 is "adding something 0 times" so we get 0. That view of multiplication isn't too problematic for whole numbers, but it gets more tricky if we're considering fractions and irrational numbers as well. What would it mean to multiply by π according to repeated addition? We can't do repeated addition π times.

As we saw in Chapter 1, mathematicians take a more general approach in order to encompass more possibilities, including not just more complicated numbers, but also shapes and other things that aren't numbers at all. The idea is to take addition and multiplication to be two separate building processes. With whole numbers we happen to be able to define multiplication in terms of addition (repeated addition) and then explore some relationships that result. So we define things like

$$2 \times 3 = 3 + 3$$

and

$$3 \times 2 = 2 + 2 + 2$$

and then we can draw a picture to discover those are the same:

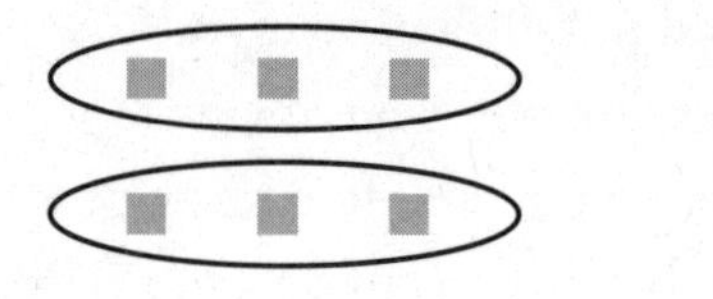

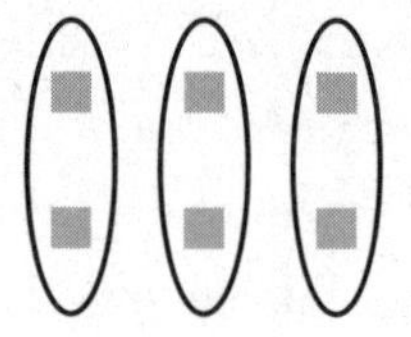

so we know

$$2 \times 3 \;=\; 3 \times 2$$

By the way, I like the fact that these pictures show that the answers are the same without us actually needing to know what the answer is. It's all about the processes, not the answers.

More profoundly we can do things like I did when explaining 6 × 8 many different ways. We could observe that 6 = 5 + 1, so

$$\begin{aligned} 6 \times 8 \;&=\; (5+1) \;\times\; 8 \\ &=\; (5 \times 8) \;+\; (1 \times 8) \end{aligned}$$

I don't want to get distracted by the order of operations here (which I'll come back to) but I just want to point out that I'm using parentheses to emphasize which things are grouped together. I would actually prefer to draw these as trees, to emphasize when we're doing the operations:

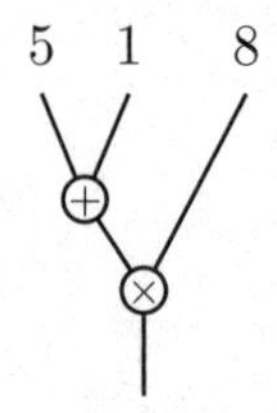

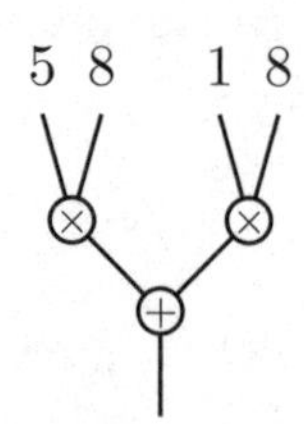

However, this diagram doesn't really give us any helpful intuition about why these two things are the same; it just helps me understand the algebra better than a string of symbols in a line. We'll come back to other ways to visualize this in the next chapter.

The convention of how we use parentheses means that, strictly speaking, some of those parentheses aren't required, so I could have written

$$5 \times 8 + 1 \times 8$$

However, that convention is just a notational convention. It's not math, it's orthography. I prefer using the extra parentheses to make things clearer, rather than insisting that everyone remember that arbitrary notational convention. More to the point, the expression above is very hard to read compared with this one:

$$(5 \times 8) + (1 \times 8)$$

Anyway this expression:

$$(5 \times 8) + (1 \times 8) = 6 \times 8$$

is something we might be able to understand abstractly, possibly better than by looking at that string of symbols. We understand deep down that if we take 5 apples and 1 apple we get 6 apples, and then we go up a level of abstraction and realize that if we take 5 "things" and 1 "thing" we get 6 of those "things," whatever those things were: they could be apples, bananas, elephants, or indeed eights. Hence 5 eights plus 1 eight is the same as 6 eights.

Then when we are trying to allow for irrational numbers we basically do all this the other way around. Instead of defining multiplication via addition and then observing this behavior, we define multiplication to be "something that behaves like this." It's a bit like seeing birds for the first time and saying "OK I'm going to call those things birds" and observing that they have feathers and fly, and then taking a step back and saying "Well, how about I call anything a bird if it has feathers and flies." And then you might decide to refine it a bit later when you discover some things that really *look* like other birds but they don't actually fly, although perhaps they look like they could *potentially* fly because they do have feathers and wings.

Those classifications sometimes caused great confusion, such as when the flying lemur was originally classified as closely related to a bat, and the pangolin closely related to an anteater. In those cases it was some superficial resemblances that temporarily led scientists astray. To this day some people who didn't pay enough attention in biology think that everything that lives in water must be a fish, and explosive arguments result on the internet with people calling other people stupid for thinking that dolphins and whales could possibly be mammals.

Characterizing numbers by behavior

We need to be careful how we characterize living creatures by their behavior, and we need to be careful with numbers too. We characterize numbers by saying: we can add them, we can multiply them, and there's some sort of interaction between addition and multiplication.

In more detail, we have a notion of addition, and it behaves like addition of blocks even when it's nothing like blocks, so:

- It doesn't matter what order we add things together, like
$$2 + 5 = 5 + 2$$
- It doesn't matter how we group things, like
$$(2 + 5) + 5 = 2 + (5 + 5)$$
- There is a number that "does nothing" for addition, called 0.
- There is a way of "canceling out" addition by any particular number, called the negative of that number.

We also have a notion of multiplication, and it does the analogous things with a tiny caveat about canceling out. The principles for multiplication are a little harder than those for addition in terms of blocks, but the analogy is quite clear when we just write it out:

- It doesn't matter what order we multiply things together, like
$$2 \times 5 = 5 \times 2$$
- It doesn't matter how we group things, like
$$(2 \times 5) \times 5 = 2 \times (5 \times 5)$$
- There is a number that "does nothing" for multiplication, called 1.
- There is a way of "canceling out" multiplication by any particular number except 0, called the reciprocal of that number.

And finally we have this principle for interaction between addition and multiplication giving things like

$$(5+1) \times 8 = (5 \times 8) + (1 \times 8)$$

and

$$8 \times (5+1) = (8 \times 5) + (8 \times 1)$$

although we don't really need to say it both ways around because we know that multiplication is the same both ways around, so the two statements are equivalent. (There are worlds where the order of multiplication does matter, and then we do have to say it both ways around.)

This last thing, about interaction, is called the "distributive law" of multiplication over addition. It can seem a bit mysterious when written out, but if we think of it as "5 things plus 1 thing is the same as 6 things," it's less mysterious. In fact, the distributive law sort of tells us that multiplication must be repeated addition *wherever that makes sense*, because for any number built from 1s, like $3 = 1 + 1 + 1$, we can deduce that 3 times anything else must be that other thing, added together three times, such as

$$\begin{aligned} 3 \times 7 &= (1+1+1) \times 7 \\ &= (1 \times 7) + (1 \times 7) + (1 \times 7) \\ &= 7+7+7 \end{aligned}$$

That might make your eyes glaze over as it's a lot of symbols in rows (it makes mine glaze over, honestly) and in the next chapter we'll see some more illuminating geometric ways of depicting this. In that chapter I'll also draw more attention to the fact that I haven't really fully stated everything above, because I've only given a few examples of how things work for particular numbers, and left you to extrapolate to all other numbers. That's too ambiguous for math, but furthermore we physically can't write down the relationship for all numbers separately, as there is an infinite number of them. That's why we use letters to represent numbers—so that we can state all of those rules at the same time, for all numbers, without having to leave anything to extrapolation. But that's the subject of Chapter 5.

For now I want to explain how multiplication by 0 follows from these basic principles, which means that the result about multiplying by 0 always giving 0 isn't a basic principle, it's a consequence of these rules. Remember, the whole reason I was doing this is to give an example of something "basic" we typically insist on proving in college math courses, which makes some people feel like we're doing things that are ridiculously obvious, which they've known all their lives.

Here's how it goes. I'm going to warn you that this might seem tedious and technical. I'm not meaning for you to understand it, but more, to gaze and marvel at its convolutedness.

We want to prove that for any number a, $0 \times a = 0$. But what *is* 0? It's the additive identity, the number that does nothing when we add it to things. So it turns out that we can investigate this situation by adding $0 \times a$ to itself. (I'm going to stop writing the × sign now, because it gets in the way visually.)

$$\begin{aligned} 0a + 0a &= (0+0)a \\ &= 0a \end{aligned}$$

But now we know we can "cancel out" $0a$, whatever it is, by using its negative, that is $-(0a)$, whatever that is. So we add that to both sides of the above equation, giving

$$-(0a) + 0a + 0a = -(0a) + 0a$$

but then each $-(0a)$ cancels out one $0a$ so all that's left is

$$0a = 0$$

This can seem off-putting to people who've taken that fact for granted as "obvious" all their lives, but for anyone who's sat around wondering *why* it's true, this can be rather satisfying. Of course, it might still not be satisfying, like when you get to the end of a book and it turns out the whole thing was a dream. All I'm trying to say is that thinking things are obvious doesn't mean you're a better mathematician. Research mathematicians just keep trying to explain things more and more. It's true that we often say things are "obvious" but there's that joke about the mathematician who goes away for a week with a problem and comes back and says "Yes, it's obvious." Really, "obvious" means "I know how to explain that."

This mathematical urge is to find deeper and deeper explanations of things, find deeper reasons why things are going on, find deeper relationships between the ideas we're thinking about. This is a very different motivation from the drive to solve specific problems in life, but sometimes the two end up coinciding, sometimes a very long time later.

Unexpectedly useful math

One of the most vivid descriptions of how useless math is comes from G. H. Hardy's much-quoted essay "A Mathematician's Apology." Hardy was an eminent mathematician at the University of Cambridge. In his essay he describes how utterly useless his research in number theory is, with what I consider to be thinly veiled pride. Unfortunately there are still some mathematicians who are disdainful of any research that is "useful," and I have issues with that because I don't think one should be deliberately and actively trying to do something with no applications. That's dangerously arrogant, and creates an unhealthy atmosphere in which people claim the moral high ground by their research being useless, and look down on people who actually directly help the world. It sounds uncomfortably similar, to me, to a rich person feeling superior because they never clean their toilet, and looking down on the people who clean toilets. (I do recognize, however, that some of the researchers in question might be saying this because deep down they're insecure about their work being so far removed from applications.)

Pursuing research for reasons other than direct applications is fine, but being actively proud of having no applications is less so. Anyway the funny part is that Hardy was working on number theory, which is a branch of math that digs much further into some of things we've been discussing about the behavior of whole numbers. Beginning with the innocuous idea of adding up the number 1 repeatedly forever, we then think about repeated addition and we call it multiplication. Then we think about building blocks with respect to multiplication, and we call them prime numbers. And then we start trying to understand which numbers are prime numbers. How can we find them? How many are there? Are there any patterns in them? How do they relate to each other?

Hardy was convinced that there would never be any applications of number theory. How wrong he was: that work is now the basis of internet cryptography, which almost all of us use every day, every time we log in to something. Our passwords have to be transmitted across the internet and need to be encoded to prevent people stealing them and breaking into our accounts. The way that is done is a rather clever use of prime numbers based on some theorems from number theory dating back to the seventeenth century. The basic principle is just that *division is difficult*. Addition isn't too hard, and multiplication as repeated addition is harder, but still possible, especially with a computer. Division, or worse, finding factors, is difficult because it is so hypothetical. For doing 3 × 5 there is a clear method, even if it is arduous. Doing 15 divided by 3 still has a clear method (by sharing things out), but what if we don't know in advance what we are trying to divide by? This is what finding factors is about: we are just trying to divide 15 by *something* to get another whole number.

You can probably recognize that 15 = 3 × 5, but if I try something bigger like 247 it's already going to be much harder to work out how we could express that as something times something else. You could just try every number in turn, but as the number in question gets bigger, that method takes more time. Crucially, the time that takes increases faster than the size of the number we're investigating, so that once the number is, say 100 digits long or 200 digits long, a computer won't be able to do it in the space of a human lifetime.

This is a clever way of making a secret, because it means that if I pick two very large prime numbers and multiply them together, I will know which numbers I used, but nobody else will ever be able to find it out. There is a rather large step in between that idea and implementing it into a workable code, but that's the basic idea. The thing that bridges the gap into a practical workable code is a theorem from the 1600s called Fermat's Little Theorem (as opposed to his big one).† It's an interesting side note that this code isn't secure *in theory*, it's only secure *in practice* because we don't have computers powerful enough to find those large prime factors in a reasonable time. One reason quantum computers would alter life so dramatically is that it is thought that they would be able to find those large prime factors in a reasonable time, which means that all internet cryptography would come crashing down and we would need a completely new way to keep online accounts secure.

† The big one is the so-called Fermat's Last Theorem, famous for the note he scribbled in a margin.

Still, the moral of this story isn't that all math is useful in the end, although that might be what it sounds like. My point of view is a little more subtle: that we often can't tell in advance what is and isn't going to be useful later, and so it is misguided to judge the value of some mathematics by what we perceive as its usefulness (or uselessness). Anything that helps us understand something has the potential to be beneficial to humans later.

Sometimes it takes even longer than the few hundred years of number theory.

The usefulness of Platonic solids

Possibly my favorite example of a long trajectory between ideas and applications involves the Platonic solids, which mathematicians of ancient Greece were thinking about over two thousand years ago. The Platonic solids are three-dimensional shapes with maximal amounts of symmetry. That's not a terribly rigorous definition but it is the general idea.

More specifically, they are three-dimensional structures built from flat two-dimensional shapes (which are then called the "faces" of the shape), with all of the following kinds of symmetry. First of all, the two-dimensional faces all need to be maximally symmetric, meaning that their angles and sides are all the same. Then the way they're stuck together needs to be maximally symmetric, so those faces all need to be the same shape and size, and moreover they need to meet each other with the same angles everywhere. Finally, the resulting shape needs to be sort of roundish. Obviously that's not a formal definition, but the idea is that there shouldn't be corners that point inward as well as corners pointing out—all the corners need to point out, so it's more like a sphere and less like a star-shape. (The technical name for this is "convex," and we'll come back to that in the last chapter.)

It turns out that there are not very many shapes we can build that fit those principles. For a start, there aren't many two-dimensional shapes we can start with that will even get us going. We could start with equilateral triangles, squares, or regular pentagons (regular means that all the sides and angles are the same). But if we try using hexagons we won't get anywhere, because hexagons fit together neatly in a plane, that is, a flat surface:

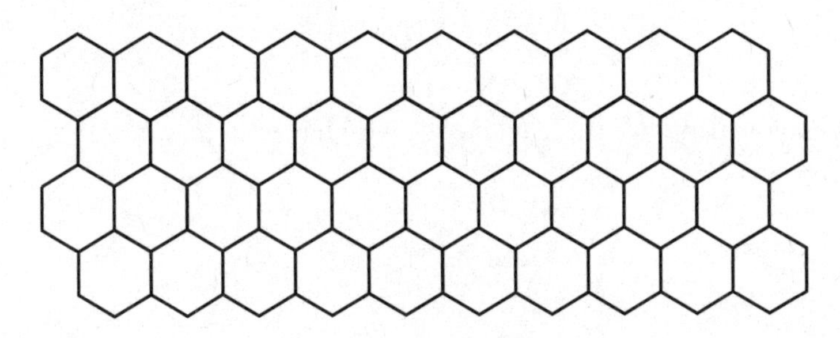

This is an interesting and indeed useful fact in its own right (and recently I've noticed graphic designers using it all over the place for company logos, carpet designs, wallpaper) but it means there's no way to build them up into something three-dimensional. Now, it's true that squares fit together as well:

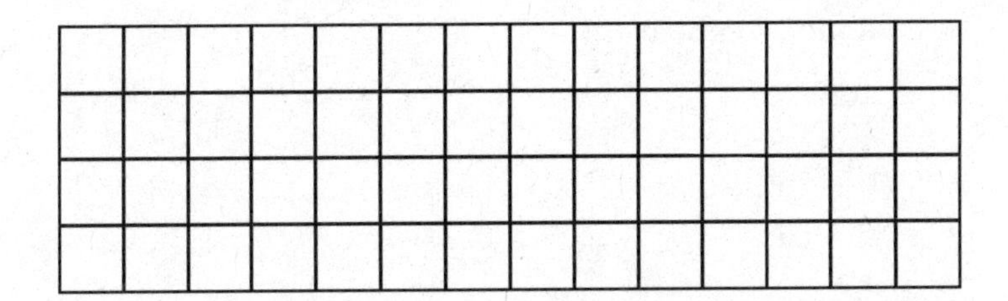

but here we have four squares meeting at each corner to fill in a flat surface, so we could just remove one. We'll then have three squares meeting at a corner, leaving a gap. If we close that gap up the thing becomes three-dimensional, and we have the makings of a cube—we just need to fill in the rest of the gaps with squares in a symmetrical way.

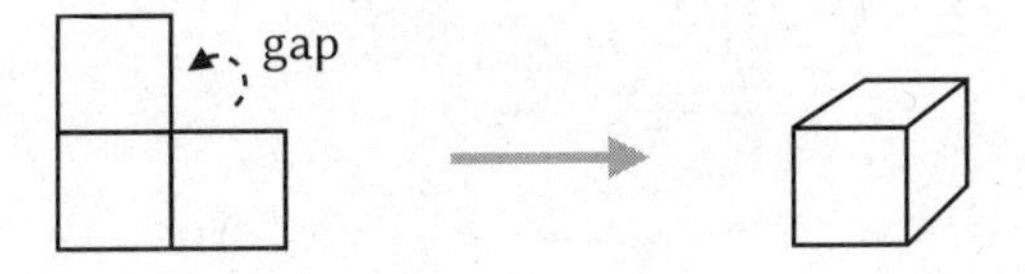

We can't do that with hexagons because three hexagons already fit together with no gap.

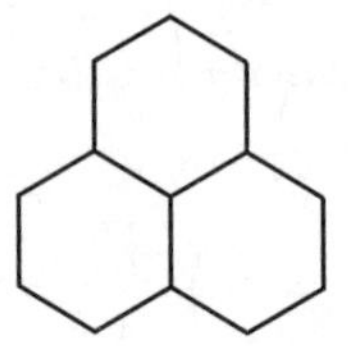

We can try this with three triangles, which leave a very large gap.

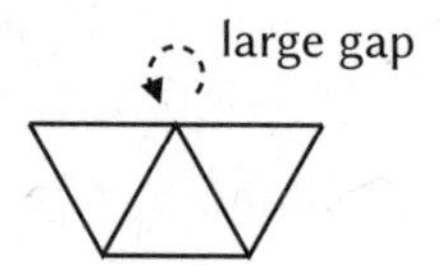

If we join those two "open" edges together we get a triangular hat-like shape, and if we stick a triangle into the last remaining space we see it's a triangular pyramid. (I recommend doing this for yourself if you have some scissors and tape around, in case you don't understand my pictures. I always find I get a better feel for things if there's something physical for me to construct.)

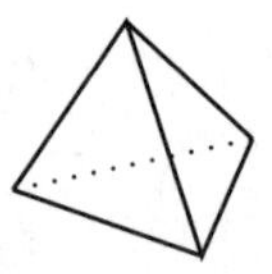

This is made from four triangles, so has been named a "tetrahedron" after the Greek for four.

Now, with the triangles, three of them left such a large gap we could also try it with four triangles, which still leaves plenty of gap for us to close up.

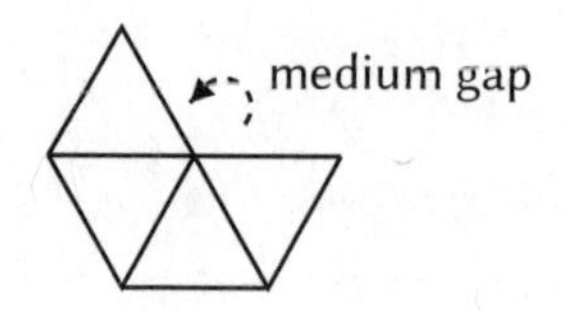

This might look like it's going to make a square-based pyramid,

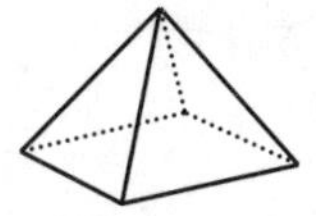

but remember we're looking for maximal symmetry, so we don't want to mix up triangles and squares in one shape (we can, but then it won't be a Platonic solid). If we keep going symmetrically we get this three-dimensional diamond-like figure:

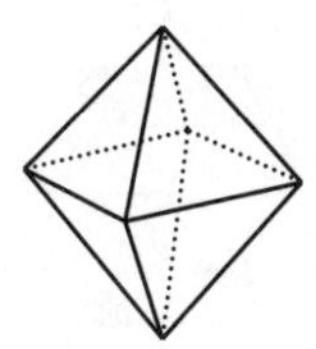

It consists of 8 triangles, so is called an octahedron.

There is one more thing we can build with triangles, because we can fit five triangles together and still leave a gap.

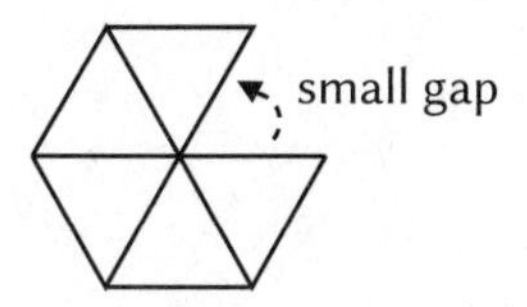

Then if we close that gap up we can make a much flatter hat-like shape:

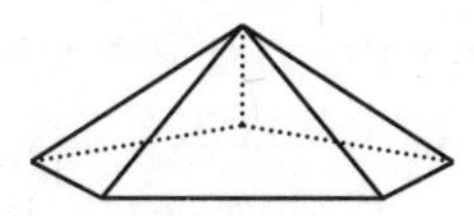

If we continue to add triangles to this shape while maintaining the pattern, it'll take longer and take many more triangles. I recommend trying this at home if you like making things. I have done it many times in my life and still find it enormously satisfying (especially if you tape it on the inside). It takes 20 triangles to close it up, so is called an icosahedron, after the Greek for twenty. Here's an attempt at a picture:

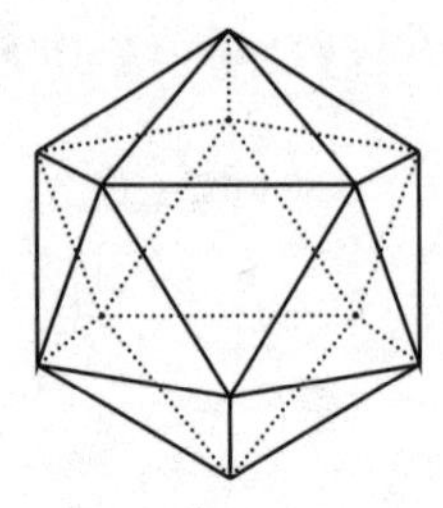

We can try this method with six triangles, but now they fit together in a two-dimensional plane, like the hexagons, so that won't get us anything three-dimensional.

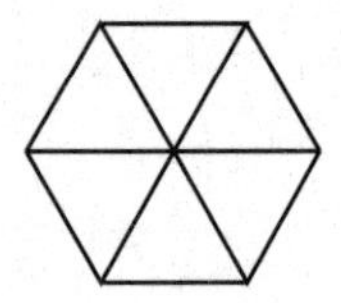

We can do it with pentagons, because if we stick three regular pentagons together there is a very small gap.

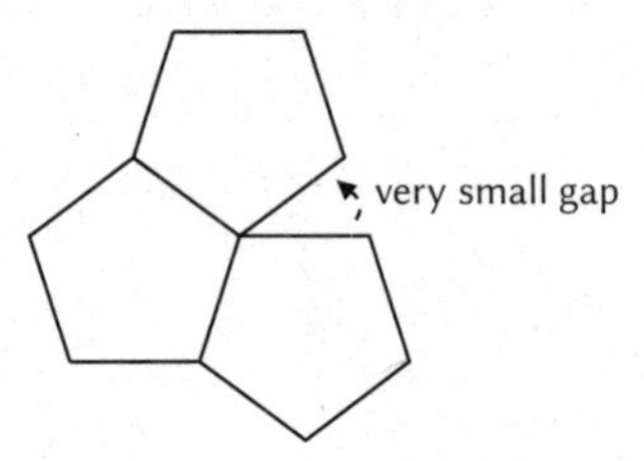

We can close that up to make a flattish sort of hat (which is a bit hard to depict convincingly in a drawing), and if we continue with more pentagons symmetrically we will close it up with twelve pentagons, so this is called a dodecahedron, and looks something like this:

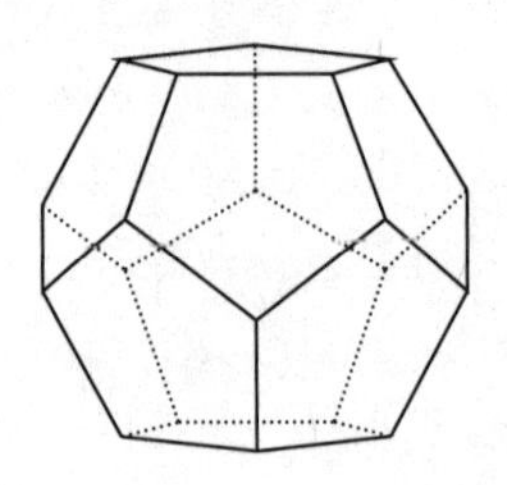

That is all the possibilities for the maximal symmetry we're looking for. We can't use any shape with more than six sides because we need to be able to fit at least three shapes together at a point, with a gap, and anything with six sides or more won't fit as the angles are too large. Note that it doesn't mean that other three-dimensional shapes are no good, it's just that we were wondering, out of curiosity, what we could make with maximal symmetry. The other three-dimensional shapes are wonderful in all sorts of ways, they just don't have this particular property.

So the Platonic solids are: the tetrahedron, octahedron, cube, dodecahedron and icosahedron.

You might wonder why a cube isn't named after the number of faces it has. Well, we could do that and it would be called a "hexahedron." Maybe that's just too much of a mouthful for something that's really quite common around us. We do often give things shorter names if we refer to them a lot. It's actually called a cube after the Greek word κύβος for a six-sided die. (We often end up with irregularities in language for things that are particularly commonplace, such as when mundane verbs behave irregularly.)

Anyway, this is all very well, but what's the point? Good question. The point, in and of itself, is just to get a deeper understanding of how symmetry works, how shapes fit together, how two-dimensional things can be built into three-dimensional things. How does this help us?

It took two thousand years for the concept of an icosahedron to become "useful." The shape is now used as a way of constructing architectural domes, in a way that was made famous by the architect Richard Buckminster Fuller (although it was first done by German engineer Walther Bauersfeld). The idea is that if we want to build something large and spherical as an architectural structure, it's hard to make it smoothly round, but if we build it from triangular pieces we can approximate a sphere from small, more convenient basic building blocks. The Platonic solids are all somewhat sphere-like, but they are also somewhat pointy. The more faces they have, the less pointy they are. What I mean is that they have more corners, but the corners are less dramatic, which is what I mean by "less pointy" at the moment. A tetrahedron (triangular pyramid) has very sharp angles, whereas an icosahedron has gentler ones.

It's still rather too pointy to be a sphere though, so Buckminster Fuller had the idea of shaving all the corners down to make the whole thing less pointy. Each corner has five triangles meeting at a point, so if you shave that corner down, it will make an extra pentagon face. And what happens to all the old triangle faces? Here's a picture of what happens when we shave off the corners of a triangle:

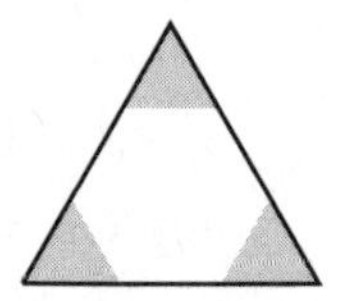

As long as we don't shave off too much, the triangle becomes a hexagon, which shows up here as the white part in the middle. So if we shave off all the corners of an icosahedron, not too far down, then every triangle face will become a hexagon, so there will now be 20 hexagon faces. But also each corner will turn into a pentagon face. An icosahedron has 12 corners, so this new shape will have 12 pentagon faces along with 20 hexagon faces and it turns out to be quite a good approximation of a sphere. It's so good, actually, that this is a widespread way of making soccer balls, and the shape is sometimes called a "buckyball" after Buckminster Fuller.

The final step to turning this into a handy structure we can build from triangles is to realize that we can make each hexagon out of six triangles, and we can make each pentagon out of five triangles. This makes the construction very convenient as you only need one type of building block: a triangle.

This is how geodesic domes are made from triangles, and they are often used for the domes of planetariums. More recently they have been used to make children's jungle gyms, and also outdoor dining bubbles, as well as eco-domes. You can spot that this is going on if you look at how many triangles meet at a point in different places in the structure. Sometimes there will be six and sometimes there will be five, showing traces of where there is really a hexagon and where there is really a pentagon face.

As a further branch of this icosahedron story, in the 1950s microscopes were made that are powerful enough to look at the structure of viruses, and many viruses turned out to have icosahedron structures. It took two thousand years for the study of icosahedra to find a use or application, and I don't think that could ever have been predicted at the time.

Using Platonic solids for outdoor dining bubbles might sound a little frivolous as applications go, but a less trivial example takes us from ancient Greece all the way to calculus.

Infinity

This is a story I've told in *Beyond Infinity* so I'll just summarize it here. People have been thinking about what infinity really means for thousands of years. The questions posed by mathematicians and philosophers thousands of years ago are almost the same questions that are often posed by curious children to this day. What is infinity? Is infinity a number? Can we ever get to infinity? Are there infinite things in the world? If we divide something up into an infinite number of pieces how big is each one?

Many of these questions were studied by the ancient Greek philosopher Zeno and his group, and the puzzling parts were encapsulated in Zeno's paradoxes. One of my favorites is the one I like to describe as "How to make your chocolate cake last forever," which is that you eat half, and then you eat half of what's left, and then half of what's left, and so on. Does that mean you will never finish your chocolate cake? Zeno didn't express it in terms of cake, but in terms of traveling from

A to B. The conclusion appears to be that you have to keep covering half the remaining distance, but that means that half the remaining distance always remains. That seems to mean you never get to B, and yet we do all arrive at places every day.

Another paradox investigates what "motion" really is. Zeno thought about an arrow flying through the air, and pondered the fact that at any given moment the arrow is only in one place in the air. So how is it moving?

Another one involves a foot race between the character of Achilles (known to be very fast) and a tortoise. The tortoise is given a head start, but of course Achilles is many times faster than the tortoise. And yet, if the tortoise starts at point A, then by the time Achilles arrives at point A, the tortoise must have got at least a little bit ahead, perhaps to point B. Then by the time Achilles has got to point B, the tortoise must still have got a little bit ahead, say to point C. And so on forever, which seems to mean that Achilles can never overtake the tortoise, but that doesn't seem to make sense at all.

There are several possible human responses to these paradoxes. One is to throw up your hands and go "This is stupid!" or "What's the point!" Another is to scoff and say "Of course you can finish your chocolate cake! Obviously Achilles overtakes the tortoise!" but that doesn't in any way address what the thought process in the paradox is actually doing, it just dismisses it.

Mathematicians do neither of those things. We feel a sense of discombobulation and confusion, but that draws us in and makes us want to work out what is going on. It took a couple of thousand years, but mathematicians finally worked out how to deal with these weird things by inventing the field of calculus, which in turn led to essentially every development of modern life, via electricity and other technology.

Confusion can be off-putting and can make you feel like you're not intelligent enough to do this, and therefore that you should go and do something else. But sometimes people who should be confused aren't as confused as they should be, which is just a form of delusion or lack of self-awareness (like when people think they can run a country despite being woefully unqualified to do so). Meanwhile other people are appropriately confused and overwhelmed because the situation really is confusing, and unfortunately they are led to believe that this is a sign that they're not cut out for mathematics. On the contrary, it's a sign that you have correctly detected that something interesting is going on, and that we have an opportunity to become more intelligent by working it out. This is what happened with Zeno's paradoxes, but it took almost two thousand years for mathematicians to work it out. So no, it's not easy, or obvious. In fact it's rather wonderful. In the next chapter, we'll explore how those conundrums about infinity led to the field of calculus, and why that's both a brilliant but also uncomfortable piece of mathematics.

CHAPTER 4

WHAT MAKES MATH GOOD

Why does 0.9 repeating equal 1? Surely it can't actually *equal* one? Is it just very, very close to 1, but it will never actually get there?

Repeating decimals are the ones that go on forever in an endlessly recurring pattern. For 0.9 repeating it's a very simple pattern:

$$0.9999999999999\cdots$$

with the three dots at the end indicating that the 9s go on "forever." It's sometimes abbreviated with a bar on top like this:† $0.\overline{9}$

Repeating fractions hold a fascination that's related to the intrigue around infinity and the idea of things being infinite. But the connection with infinity also means that recurring decimals can be rather confusing, and that our intuition can lead us astray.

This can result in people with different intuition getting into arguments about what is "right" and what is "wrong." In this chapter I'm going to explore one of the important ways in which math isn't just about right and wrong. We've already seen that math has

† The notation is different around the world; for example some countries including the UK use a dot over the 9 instead of a bar.

a strong framework for deciding whether something counts as true, but there's much more to math than that. Even if a piece of math is logically sound, we still evaluate how good it is. This is much more subjective than logical correctness, and might include matters of taste such as whether or not you find it helpful, illuminating, or satisfying. I'm going to explore what sorts of things are valued in mathematics, including how math interacts with our intuition (either backing it up or fixing it when it's faulty), how good math sheds light rather than just saying something is true or false, and how the shedding of light also helps us to unify more situations and make our arguments more broadly applicable. I'll talk about how math enables us to build more and more complex thoughts and make progress, but we must also acknowledge that this is just a value system that has been developed by the field of academic mathematics as built by a culture of European white men. This raises some uncomfortable questions about why we value these things.

Mathematical values

What makes math good is a much more nebulous question than whether it's right or wrong. And it's a much more interesting question. Mathematicians don't all agree about it all the time, but there is often a broad consensus about certain things within fields, perhaps just with a few outlying dissenters. This happens when we make judgments about anything in life, whether it's math or movies. At the time of this writing, the highest rated movie on IMDb is *The Shawshank Redemption*, and I can imagine that there is broad consensus that this is a great movie, even if it's not to your taste (I love the ending, but there's too much violence in the middle for me to be able to watch all of it). The next is *The Godfather*, which I also can't watch because of the violence.

Mathematical beauty is a contentious issue, and I have had people object to it on the grounds that mathematicians can't agree on it or define it. I don't buy this as an objection though, because we humans don't agree about any kind of beauty, nor have I ever seen a clear definition of it. Personally I've given up on the idea of physical beauty (for people), as I genuinely think that beauty is only about kindness and generosity, and I can no longer bring myself to care what anyone looks like physically. I admit that in part this is my defiant response to what I perceive as the superficial beauty culture and especially its pressures on women.

Math is not all about right and wrong, although there is a concept of right and wrong because there is a concept of logical contradiction. But as I discussed in Chapter 2, that notion of right and wrong operates at a more abstract level than the one people are usually thinking of when they think of math as having right and wrong answers. For example, "What is 1 + 1?" doesn't just have the correct answer 2. As we've seen, it has many different possible correct answers in different contexts.

However, "If 1 + 1 = 1, what is (1 + 1) + 1?" does have a correct answer, because it's a question of logic.

"In the context of the natural numbers, what is 1 + 1?" does have a correct answer, because it's a question of standard mathematical definitions in a specific context.

I would argue that this is a level of "correct answer" that exists everywhere, even in fields that are not usually thought of as having correct answers. For example, even in art there are correct answers. If you mix two colors of paint together there is a particular color that will result, and you can't change that: it just happens. You can't just choose for it to be something else. You can choose to mix the colors in different proportions, or you can choose to mix different colors, but once you've done it, there is an "answer" of what color results. The only way to get a different answer is to use a different process.

If you're building a structure it will either stay up or it will fall down. If you're making a dress then there are many ways to do it, but if you want it actually to stay on someone's body then there are some things you need to do to make it stay there, or it will fall off. (You might want it to fall off, but in that case there are things you need to do to make it fall off, otherwise it will stay on.)

The "rules" for writing have become gradually more and more expansive over history. The rules of grammar have been changing, and writers push the boundaries of what counts as a sentence, what counts as a word, and even how words are spelled. Every generation has people who object to these things, which they consider to be "wrong," but language has always been evolving. In the case of writing, aside from the question of what is "correct" and "incorrect," and what is "good" or "bad," there is also this question: Is anyone going to want to read it? That said, at some very purist artistic level a writer might want to write exactly what they feel like, with no externally imposed constraints at all, and without any care for whether or not anyone ever reads it. Such a writer might follow absolutely nobody's rules but their own. Here's a short poem I have written to express the pain of traumatic grief:

again
aghghast

no no no

But even if you want to pursue a creative process that imposes no rules at all, there is still value to training our brains to process logic that entails a concept of right and wrong, because *life* has that in it. If you believe that all immigrants are illegal that is logically incorrect. If you believe that a vaccine gives you 100 percent protection against an illness then that is logically incorrect; if you think it gives you no protection then that is also logically incorrect.

If you think that science claims climate change is certain then that is logically incorrect. If you think that means climate change is definitely not real then that is also logically incorrect. These things do come up in life and I wish everyone had stronger core brain muscles so that we could all keep the logic of various arguments clear. I'm not saying that everyone who disagrees with me is illogical though: there are logical ways to disagree. It's not logical to think that vaccines give no protection against illness, but it is valid to decide you fear the side effects of the vaccine more than you fear the illness. It is a different response to risk from mine, but at least we can then discuss that difference.

That was all a defense for the aspect of math that does have a concept of right and wrong, but now I'd like to focus on more subtle nuances of why we value math, and what counts as good math. Although mathematicians don't all agree on what makes for good math, I'd like to say a few things about what some of us think, including me. We'll start with the example of 0.9 repeating, and see that this question led to a clever (albeit slightly sneaky) way to pin something down that our intuition might have trouble with. Later we'll look at situations that are more about helping our intuition along, or even building arguments from our intuition when we believe it is pointing the right way.

In life I believe it's important to remain open to new information whether or not it shows that our intuition is pointing us the wrong way. I believe this in math too. With repeating fractions, as with many questions about infinity, our intuition can fall short, because we just don't have a lot of experience with infinity in our short, finite lives. I enjoy the fact that some rigorous thinking can "correct" our intuition if we are open to it, but the process of admitting that your intuition was flawed and allowing it to change direction can be unsettling. It's a bit like when our unconscious biases about people are shown to be flawed. In both cases we can either cling to our preconceived ideas, or celebrate our capacity for honing our thinking. Mathematics does the latter, and I try to as well. The question of 0.9 repeating is a good place to practice it.

What is a repeating decimal really?

In general, irrational numbers are "decimals that go on forever without repeating," and as we don't have forever it's very hard to say what they are unless we can characterize them some other way. For example $\sqrt{2}$ is "a number whose square is 2," and π is "the ratio between any circle's circumference and its diameter" (which we'll come back to in Chapter 6). But repeating decimals are the ones that go on forever *with* a repeating pattern, so we do know what they're going to do "forever," at least step by step. The trouble is that this doesn't tell us what the result is; perhaps it's like knowing what repeating pattern a spiral staircase takes but not knowing where it takes us.

We can try thinking about "0.9 repeating" in steps. The first step is 0.9, which is $\frac{9}{10}$. We know that this is quite a large part of 1, but definitely not all of it. We could represent it in this picture:

The next step is 0.99. This is $\frac{9}{10} + \frac{9}{100}$. Another way of thinking of it is we take $\frac{9}{10}$ and then we add on $\frac{9}{10}$ of what's left over, a bit like the scenario of eating half of your chocolate cake and then half of what's left. Here's a picture:

Now we keep going. 0.999 is where we add on $\frac{9}{10}$ of what's left again.

It has already become rather hard to see the tiny gap left, but there really is a short vertical sliver of a gap left over at the top right. Here's a zoomed-in picture of that corner:

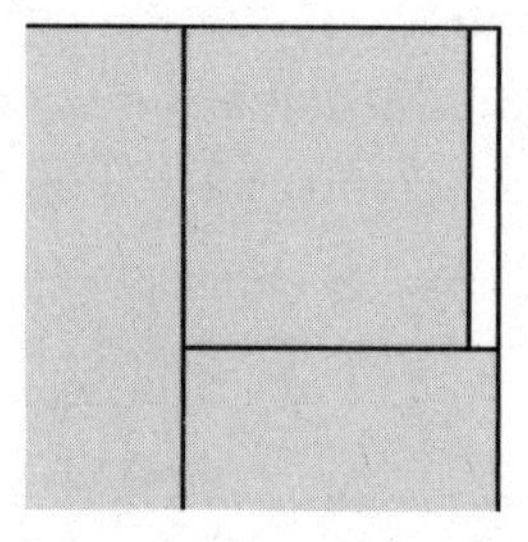

The fact that we couldn't see much until zooming in is already a sign that this argument is not very rigorous.

The idea of 0.9 repeating is that we "keep doing this forever." If you imagine continuing the pattern of the picture forever, it does look like it's trying to fill in the entire square. But here's an even more zoomed-in picture of the corner, with a few more iterations filled in, and a tiny gap still just about visible:

Some people think that you will always be able to zoom in on that corner and find a little gap, so it will never actually reach 1. Others will argue that there's only a gap left because we're not yet at infinity, and it will reach 1 "at infinity" so the answer is 1.

It's tempting to get into a fight about who is right in this case. That's a tendency ingrained in us by a society that pushes us toward zero-sum games where there has to be a winner and a loser, which means that in an argument someone has to be right and someone has to be wrong. I prefer to think about the *sense in which* everyone has a point.

If you think there is always going to be a little gap left, you are right in the sense that we humans can never actually get to infinity in this process, and so at any point that we humans can draw the situation, there will be a tiny gap left.

On the other hand, if you think that the shape will be all filled in "at infinity" that is more or less the point of this situation, but it takes quite a lot of work to make sense of that rigorously, and that's the point of calculus. Mathematicians had been trying to make these arguments about what happens "at infinity," and the arguments seem to make sense up to a certain point, but when you start really asking questions about it, and not just accepting the answers, you find that it's all on shaky ground unless you sit down and really pin down what some things mean.

Another popular argument to show that $0.\overline{9} = 1$ goes a bit like this: $0.\overline{9}$ is $0.9999999999999\cdots$. If we multiply that by 10 we know we just move the decimal point one place to the right, so we get $9.99999999\cdots$. But this is the same as $9 + 0.9999999999999\cdots$. Now let's give $0.9999999999999\cdots$ a letter name, say x, for the sake of this argument. What we've shown is that

$$10x = 9 + x$$

and now it's a case of solving that equation. If we subtract x from both sides we get

$$9x = 9$$

and thus $x = 1$.

This argument is along the right lines, but it has, I would say, a technical problem and an emotional problem. You might never have thought about the possibility of an emotional objection to math, but I emotionally object to a proof if it doesn't explain or illuminate the situation to me. I might have to concede that its argument is logically correct, but that only means it tells me the answer is correct, it doesn't help me understand *why* the answer is correct, and that leaves me unsatisfied. The above argument feels like a slick piece of trickery to me. For some people that is very satisfying, but I don't like trickery. I like transparency and clarity.

The more serious problem is a technical problem though: the above argument isn't just emotionally unsatisfactory, it's also got some giant logical holes in it. For a start, how do we know that we can multiply 0.9999999999999··· by 10 by moving the decimal point to the right? How would we justify this logically? Well in fact this pushes us to ask the fundamental question: what does 0.9999999999999··· even mean in the first place? If we don't know what it means in the first place, we can't make arguments with it. The above argument is correct, but only once we've made some definitions and proved some quite deep theorems in calculus, which are hidden in that innocuous little step about "moving the decimal point to the right." This is what my math teacher at school used to call "cracking an egg with a sledgehammer": using some really rather excessive tools to do something much simpler. In this case it's not just that the tools are excessive: the whole approach is disingenuous, because the amount of work that goes into proving the deep theorems in calculus is far greater than the work that goes into proving that $0.\overline{9} = 1$. It's a bit like proving you can climb the first step of a staircase by going down from the top: not only have you not proved you can get to the top in the first place, but getting to the top would probably involve climbing the first step directly.

And it's not just about effort—if you understand enough to understand the proofs of the deep theorems, then you've understood enough to get at why $0.\overline{9} = 1$ directly. And more to the point (for me anyway) the direct argument uses and illustrates a really brilliant and fascinating idea in mathematics, an idea that set off one of the most important developments in the modern world: the development of calculus.

The beginnings of calculus

Calculus grew, over thousands of years, from the urge to understand infinitely small things. This is related to the urge to understand continuous motion, and curves. A curve is a rather extraordinary thing if you think about it. It's made up of tiny dots that are connected to each other, but it's always changing direction. If it kept going in the same direction it would be a straight line. If it even kept going in the same direction for more than an instant then it would have a straight part, and not be smoothly curved.

We might try to approximate a curve by a series of straight lines, and the ancient ways of approximating circles are by fitting a polygon (that's a shape made from straight edges) inside or outside them. In the following pictures I've fit squares inside and outside the circle, and then octagons.

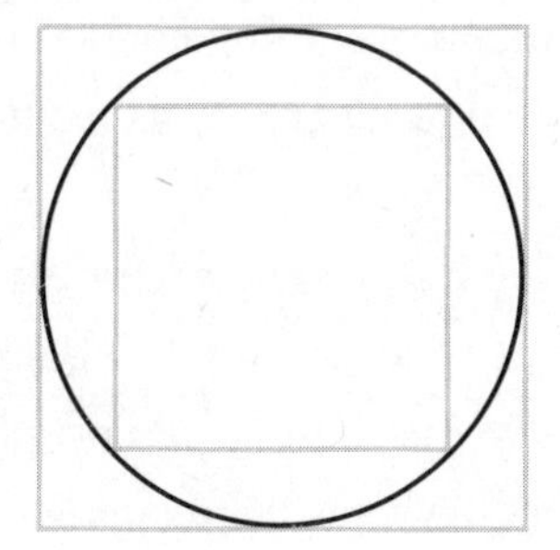

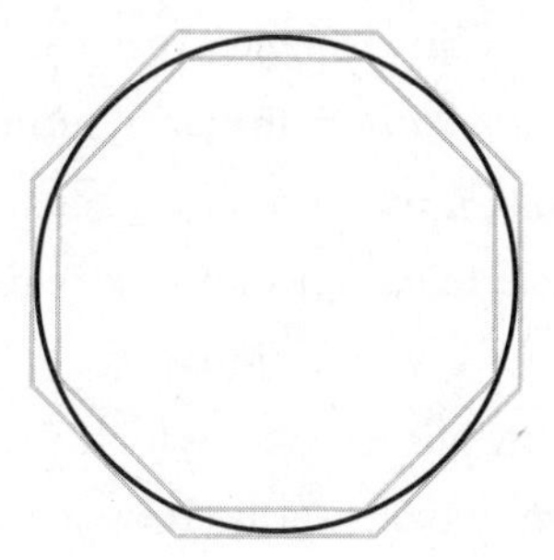

The two octagons are much closer together than the two squares, but there's still a noticeable gap between them where the circle sits.

The idea is that the more sides you use, the closer together your inner and outer approximations become, and you sort of pin the circle down by sandwiching it in between those things. It's a bit like when you have a pile of curvy pastrami between two pieces of bread and you press it down in order to get the bread closer together to fit the sandwich in your mouth. (There is actually a result known as the "Sandwich Lemma" in calculus that basically encapsulates this idea.)

Zeno was trying to understand motion by thinking about it moment by moment, which means dividing time up into infinitely small pieces. How can we add those together? A finite length of time can be divided up into infinitely small pieces, but there will be infinitely many of them. How can we add up infinitely many infinitely small pieces? In the case of the child and the chocolate cake, if they keep eating half of their remaining cake, those pieces will become smaller and smaller, and if we keep going "to infinity" there will be infinitely many pieces. How can we add those up?

In the case of 0.9 repeating, we are doing the same thing: we are trying to add up infinitely many fractions that are getting infinitely small. You might think this is just part of addition of numbers, but in ordinary numbers we've only thought about how to add two numbers. We can iterate, by adding the result to another number, in which case we've effectively added three numbers, and then we can add that result to another number, so we've effectively added four numbers, and so on. This is called proceeding "by induction," and it enables us to add any *finite* number of numbers, but it doesn't enable us to add an infinite number of numbers.

Incidentally this is mathematical induction, which is logically rigorous. That's different from the concept of an argument by induction in philosophy, which is not logically rigorous. In philosophy it means extrapolating from a finite number of events, for example, "the sun has risen every morning of my life thus far so it will rise tomorrow." That is not a logically sound argument: even though the conclusion is correct, the logic is not rigorous. Just because something has happened every morning of your life so far, that does not mean it will happen tomorrow. It's a slightly subtle point, because the fact that the sun has risen every morning so far is an indication of something going on with the laws of physics, and the laws of physics ensure that the sun will rise tomorrow. However, this does not mean that the previous sunrises are logically causing the future ones.

On the other hand mathematical induction is rigorous. It says: if we can prove that something is true for the case $n = 1$, and if we can also prove that being true for the nth step logically implies it being true for the $(n + 1)$th step, then it must be true for every finite step. The difference is that this form of induction involves a logical proof that the next step follows from the previous one, rather than just an observation that the result is true all the way up to n.

So mathematicians sat down to try and work out how to add up an infinite number of numbers. The first thing to note is that if the numbers aren't getting smaller and smaller, then we're not going to be able to do it, because the sum will keep getting bigger and bigger "forever." You might want to say that it adds up to infinity, but we can't really say that, because infinity isn't a number. (Instead we say that the sum doesn't converge.) So we can only add up an infinite number of numbers if it's a sequence of numbers that's getting smaller and smaller. Well, they could oscillate a bit, but they need to be getting smaller overall.

The eventual definition is really rather clever. It's important to note that it's just a definition. It's something mathematicians have come up with that enables us to reason with these things in a way that is consistent and productive. That doesn't mean it's correct in any absolute sense. But I do think it means it's rather good, and that's what we're thinking about in this chapter: what makes math good, rather than what makes it correct.

The idea is that rather than defining an actual process of adding an infinite number of numbers, we're going to just say how to check whether a proposed answer is a good candidate. This neatly sidesteps the question of *how* we add an infinite number of numbers. I find this a brilliant and satisfying way to reason with things that are otherwise not precise enough to be reasoned with, but I can also see how it might seem unsatisfying as we haven't answered the question head on. The thing is, some things can't be answered head on, and attempting to do so ends up being even more unsatisfying.

So instead, in the early nineteenth century, mathematician Bernhard Bolzano came up with the idea of the "limit" of a sequence, which is essentially a number that is a good candidate for the "infinite sum." The idea is that we can't genuinely keep adding numbers forever, but we can imagine going to any finite point in that adding process and seeing how close we are to a particular number. If we can get to within any tiny distance we could ever think of, then that number is a good candidate to count as the infinite sum.

Now, there might not be any good candidates at all; for example if we keep adding the number 1 forever the result will just keep getting bigger and bigger, and so will not narrow down to one particular number. However, we can prove that *if* there is any good candidate, then there can only be one. The technical name in math is "limit."

Now here's the clever part: we *define* 0.9999999999999··· to be a limit. It's the limit of the sequence 0.9, 0.99, 0.999, 0.9999, ···, the number that this sequence of numbers homes in on.

And that limit is 1.

So it's true that the sequence of finite truncations 0.9, 0.99, 0.999, ··· gets closer and closer to 1 without ever reaching it. But we define $0.\overline{9}$ to be the limit of that sequence, and the limit of that sequence actually is 1. That's the number that the definition of limit gives us.

It might sound like I've somehow redefined the notion of equality, but I haven't. It doesn't come down to the definition of equality, it comes down to the definition of limit. In summary it comes down to the following two points:

- $0.\overline{9}$ is *defined* to be the limit of the sequence 0.9, 0.99, 0.999, ⋯.
- That limit is genuinely equal to 1.

You are welcome to object to this, and I personally think it's a wonderful idea to object to math that doesn't make sense to you. I always ask students how they feel about the math that we're studying, and they're often taken aback because they've never been encouraged to have feelings and opinions about math. If you don't like the argument about $0.\overline{9}$, that's fine, and you're perfectly entitled not to like it. But if you want to object to it logically, there are only a few ways to do that: if you think $0.\overline{9}$ isn't exactly 1, what is it? There are two other options: either you think it's some other number, or you think it isn't a number. If you think it's some other number, what number is that? It wouldn't make sense for it to be less than one, because the sequence of truncations will eventually pass that and get closer to 1 than to your number. It wouldn't make sense for it to be more than one either, because the sequence of truncations definitely never gets bigger than 1, which means that the sequence will always be closer to 1 than to your number.

If you think that $0.\overline{9}$ shouldn't be a number, that's more of an interesting philosophical question. We have made a logical theory that enables us to define this to be a number. According to that theory, the number must be 1. There are no logical contradictions to it. But moreover, the ideas behind the definition of limit enabled mathematicians to define the entire real numbers, that is, the rational numbers (fractions) and the irrational numbers together. This in turn enabled them to study functions that are continuously changing, building from the concept of a limit into the entire of calculus. Calculus, in turn, has enabled us to construct most of the modern world. So, not only does this theory not have any logical contradictions, it has had sweeping, far-reaching, world-changing effects.

I see that the logical steps might be difficult to swallow if they don't seem to match your intuition about the sequence 0.9, 0.99, 0.999, ⋯

never *really* reaching 1. But the thing is that the logical steps of the definition of $0.\overline{9}$ remain correct whether or not they match your intuition. Math stands or falls on whether or not it has logical contradictions, not on whether or not any individuals can get it to line up with their intuition.

I can see why this might seem frustrating, if you like using your intuition about things. But when there's a mismatch between intuition and logic, a mathematical urge is to sit down and try to understand why your intuition isn't matching up with the logic. Then we either find a flaw in our logic which we can fix, or we have a chance to improve our intuition.†

Sometimes we might just start out with very little intuition for something, and then a mathematical argument can help us develop our intuition, by shedding light on the situation rather than just proving it.

Shedding light

I like math that sheds light on *why* something is going on, rather than just proving that it *is* going on. One of my favorites is this way of multiplying two two-digit numbers together, say 18 and 24. We write 18 as 10 + 8 and we write 24 as 20 + 4 and then we imagine multiplying them in a grid, with 24 rows of 18, or 18 columns of 24. It's a bit long-winded to draw all of those rows and columns though, so we can just abstractly sketch them like this:

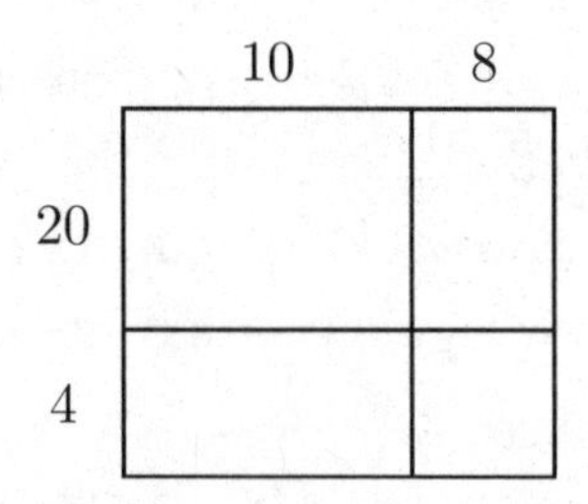

† David Bessis writes about this in *Mathematica*.

Note that it doesn't matter whether or not this diagram is to scale (mine is not). It's a "schematic" diagram showing a scheme of interactions rather than a direct representation. The idea is to invoke our geometric intuition about *interactions*, not about actual size and shape. It's more evocative than the notation we tried to use with trees in the previous chapter, which was good for representing different ways of grouping things together, but didn't invoke any further geometric intuition about the actual interactions.

With the grid representation, we can now work out how many things are in the individual boxes: each one is a rectangle, so we multiply the two edge numbers together as if we really had that number of rows and columns. The results are shown here:

	10	8
20	200	160
4	40	32

Finally we add them all together to make 432.

In a way this is not quite a piece of math as much as a method for doing multiplication, but it illustrates what I mean by "shedding light," which is also why I've decided to call it a scheme rather than an algorithm. It's a subtle point, because it does go through all the same steps as old-fashioned long multiplication, in which we do something like this (perhaps in some other order):

$$\begin{array}{r} 2\ 4 \\ \times\ 1\ 8 \\ \hline 2\ 4\ 0 \\ 1\ {}^{1}9\ {}^{3}2 \\ \hline \mathbf{4}\ {}^{1}\mathbf{3}\ \mathbf{2} \end{array}$$

We might (or might not) get the right answer, but we don't necessarily get any insight into why we're doing anything. Like with column addition, I would call it an algorithm that enables us to disengage our brain, which might be a good thing, but it also means we can bypass our intuition.

The method with the box diagram enables us to disengage our brain a little, but does so by means of relying more on our geometric and visual intuition, which I like. Later we'll see another pleasing aspect of this method, which is that it generalizes well to other situations.

I will concede that this box thing is not exactly what I'd consider to be deep math; it's more of a helpful way of visualizing something. One helpful method that I do think contains deep math is a trick my mother taught me when I was quite young, for telling whether or not a number is divisible by 9. For the numbers up to 90, you can just add the digits together and they should always equal 9. A more visual way of seeing this is to mark them on a grid of numbers, like this:

0	1	2	3	4	5	6	7	8	9
10	11	12	13	14	15	16	17	18	19
20	21	22	23	24	25	26	27	28	29
30	31	32	33	34	35	36	37	38	39
40	41	42	43	44	45	46	47	48	49
50	51	52	53	54	55	56	57	58	59
60	61	62	63	64	65	66	67	68	69
70	71	72	73	74	75	76	77	78	79
80	81	82	83	84	85	86	87	88	89
90	91	92	93	94	95	96	97	98	99

Then we can start to get a sense that the reason this pattern happens is that counting up by 9 on the grid is the same as going down one place (which adds 10) and then back one place (which subtracts 1).

This means we're always adding 1 to the first digit and subtracting 1 from the second digit, so the sum stays at 9 the whole time.

There's something satisfying to me about how this pattern fits together to produce the diagonal of the grid of numbers, but I count this as deep math because of how and, more to the point, *why* it extends to numbers of any size: for a number to be divisible by 9, you can add the digits and see if that sum is in turn divisible by 9. If you still can't tell, you can add the digits again, and keep going until you get 9 as above (or not). For example if I start with 95,238 I can add the digits and get $9 + 5 + 2 + 3 + 8 = 27$, and I can add the digits again to get $2 + 7 = 9$, so the original number was divisible by 9.

Of course these days we're rarely far from a calculator (on a phone or computer) so we could just type the division calculation in and see if the calculator gives us a whole number answer or not. And that's even aside from why we'd ever want to know if something is divisible by 9 in the first place, without just needing to know the answer. I agree that this particular piece of math has no directly useful purpose (or, as the wonderful category theorist Richard Garner says, I can think of "no useful use for using it"). For me it's definitely more in the realm of indirect usefulness, that is, shedding light on how things work. I could contrive some situations in which you might need to know if a number is divisible by 9 but, just like with contrived math homework questions involving implausible numbers of watermelons or wild horses, I don't think that contriving such a situation helps, partly because it would be obviously contrived, and partly because it would detract from the fact that it really is the indirect usefulness that is the point here. I don't want to detract from that by faking a direct use.

Explaining why this "trick" works is a bit long-winded without using mathematical notation, but the ideas involve the profound mathematical principles of how divisibility works, and how place value works. But even the idea for just the two-digit numbers and the diagonal of the number grid is very satisfying to me, as I see the ideas fitting together in a pleasing, abstract jigsaw puzzle.

Abstract jigsaw puzzles

That sense of things fitting together with nothing sticking out in a weird way is one aspect of mathematical "beauty" to me. If you enjoy the feeling of fitting the last piece into a jigsaw to produce a complete picture, that's a bit like the feeling I get from good abstract math.

The geometric image of multiples of 9 on a diagonal is an example of that to me. Another is the depiction of the factors of 30 that I like to show, where you don't just write them down, but you also arrange them to show which ones are factors of each other, whereupon they form this cube:

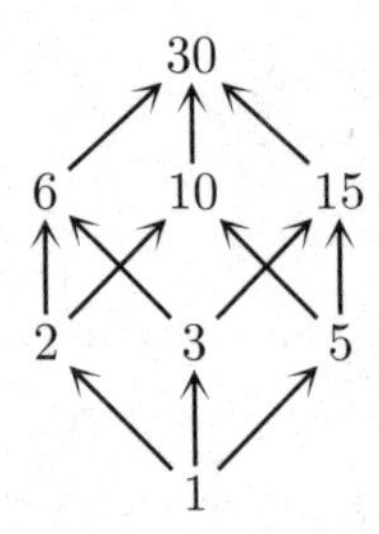

It feels very satisfying to me that the numbers fit together in this familiar shape. There are further satisfying jigsaw-like aspects to this depiction. One is that if we take it seriously as a three-dimensional cube, each dimension represents one of the three prime factors of 30, which are 2, 3, and 5. Then all parallel arrows represent multiplication by the same prime factor.

Moreover, we can see square faces representing the factors that are products of two prime numbers, that is, 6, 10, 15:

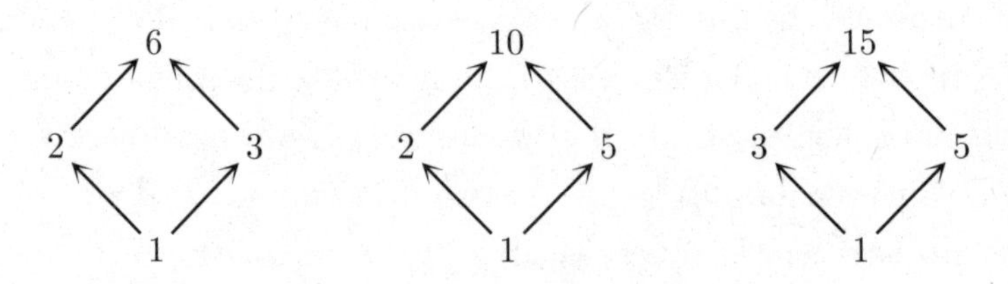

If we pick one of those faces and look at the face opposite it (really imagining it to be a real 3-D cube) we might be able to see a relationship: the opposite face is obtained by multiplying the first square by the remaining prime factor. For example we could take the square for 6, which involves the prime factors 2 and 3. The remaining prime factor is 5, and if we take the square for 6 and multiply each corner by 5 we get the opposite face of the cube as shown here:

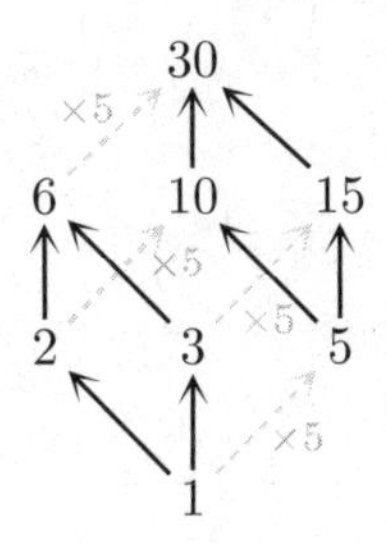

This idea can help us fit other "jigsaw puzzles" of factors together, for example if we take the square for 6 and multiply the whole thing by 2 (corner by corner), we're not going in a new direction, because 2 is already in that square. So we get this, which is a diagram of factors of 12:

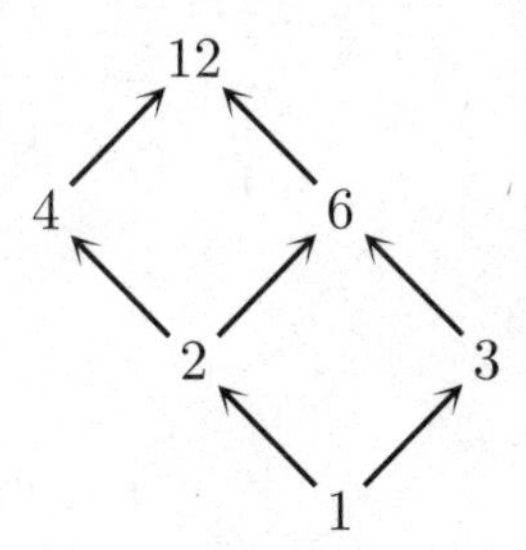

This sheds light on the interaction of factors of numbers, feels like a satisfying jigsaw puzzle, and also has another feature I like, which is that it can be generalized to more situations, including more numbers, and also expanded to unite a wide variety of apparently very different examples.

Generalization and unification

I love math for its ability to expand its scope, to encompass and unify different situations. One aspect of this (which I mentioned in the previous chapter) is *generalization*, where we expand the theory to allow more examples in, by being a little less specific in what we're talking about. Generalization in this sense is about making things more general, the opposite of making things very specific.

The method for testing if a number is divisible by 9 can be generalized not only to bigger numbers, but for divisibility by 3. Our diagram of factors of 30 can also be generalized to different numbers. We can make an analogous diagram for the factors of 42, which looks like this:

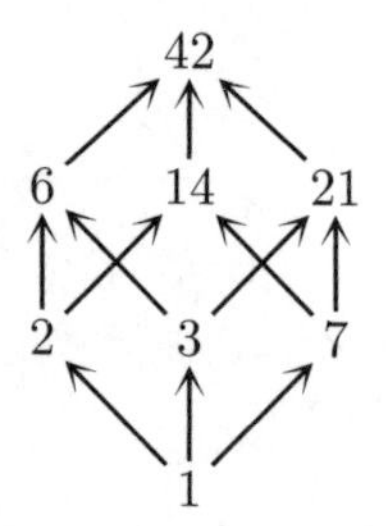

However, not all numbers produce a cube. The factors of 24 will produce this diagram:

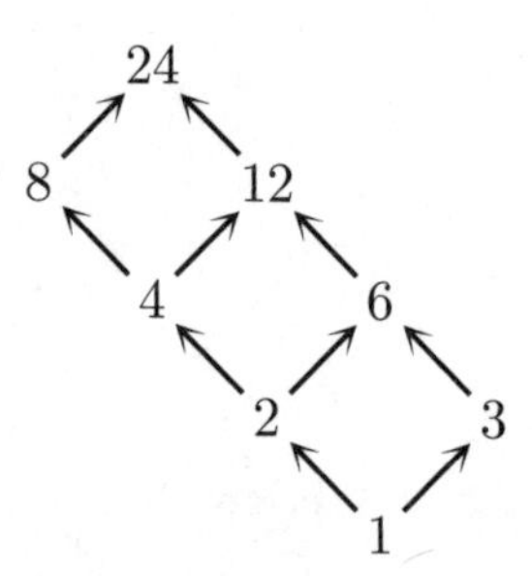

Remember, the idea is that we are highlighting which numbers are also factors of each other, by drawing arrows between numbers and their factors. But like in a family tree, we don't draw "redundant" arrows across two generations, as those can be deduced.

The method of testing for divisibility by 9 is quite tied to numbers, and it might seem that these diagrams of factors are indelibly tied to numbers too, but that is not the case if we dig deeper into *why* these diagrams are arising. This takes us back to the question of prime numbers as building blocks for multiplication, and it turns out these diagrams are exhibiting all the ways of building all the factors in question, using our multiplicative prime number building blocks.

For example for 30, the three building blocks we have in play are 2, 3, and 5. The prime factorization of 30 involves one of each of those blocks, as $30 = 2 \times 3 \times 5$. The diagram shows all the things we can build using at most one of each of those blocks.

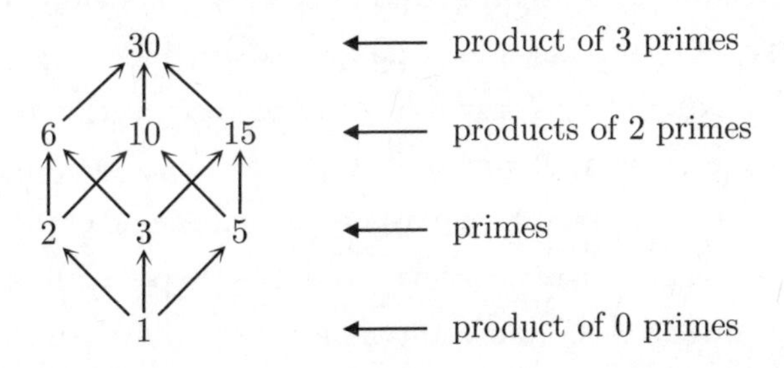

We start at the bottom with the multiplicative identity, 1, which is where we haven't done anything yet; this is why I've called it a "product of 0 primes." At the first level there's everything we can do with one block, at the next level there's everything we can do with two blocks, and at the top there's the one thing we can do with three blocks, which is multiply them all together to get 30.

This shows why we get the same diagram whenever we start with three different blocks. For 42 we are starting with 2, 3, and 7. The number 24 is different because its prime factorization is $2 \times 2 \times 2 \times 3$ so this time we are starting with *three* of the same block 2 and just one of the block 3. This creates a different interaction of possible things we can build, because when we're using two blocks we could use two of the same thing and get 2×2, or we could use two different things and get 2×3. But we still have a hierarchy according to how many "blocks" are used to make each number:

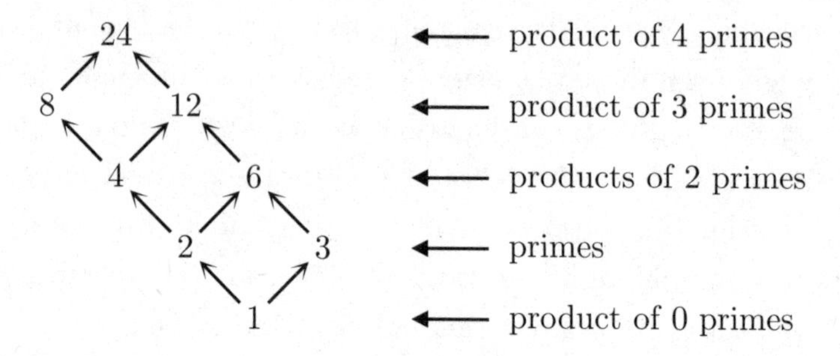

So far we've generalized this situation to different numbers, but understanding it at this level means we could generalize it to any building blocks and any method of building. I have previously written about how we can apply it to privilege and look at the interaction between any three types of privilege, for example rich, white and male. The method of building in this case is just gaining privilege, and we get a diagram as shown below, which I often talk about as I find it so compelling. At the bottom level we have people with none of those three types of privilege. At the next level we have people with one type, so there are three possibilities. At the next level we have people with two types, so there are also three possibilities, and various arrows between those levels showing the gain of one type of privilege. Then at the top we have people with all three types of privilege.

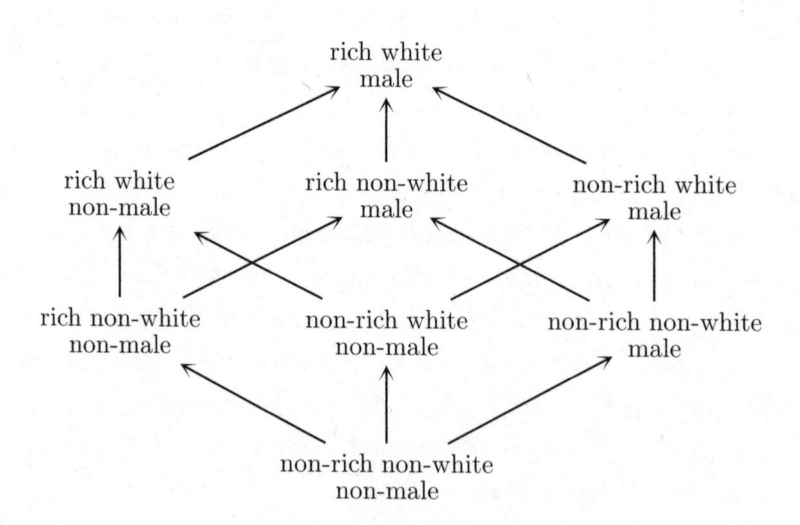

But we could extend this to an analogy with 24, using the idea of repeating a building block. Thus if we gain the privilege of wealth several times, we could think of richer and richer people. We could consider poor people, comfortably-off people, rich people, and super-rich people, and get a diagram analogous to the one for 24 like this:

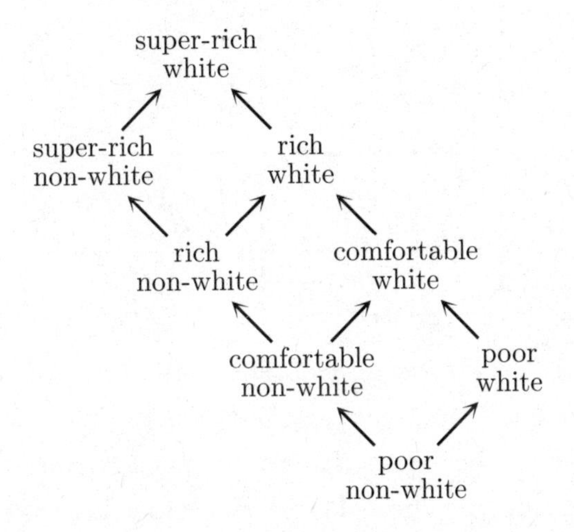

The way that this form of diagram has united some very disparate situations is a very powerful aspect of math to me. It's not just that it is applicable, which is like a one-dimensional move:

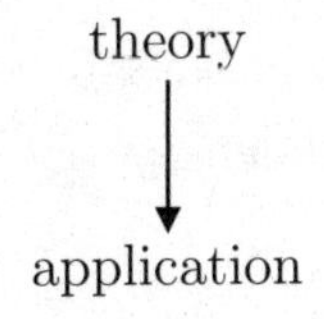

It's that it has unified a wide variety of situations:

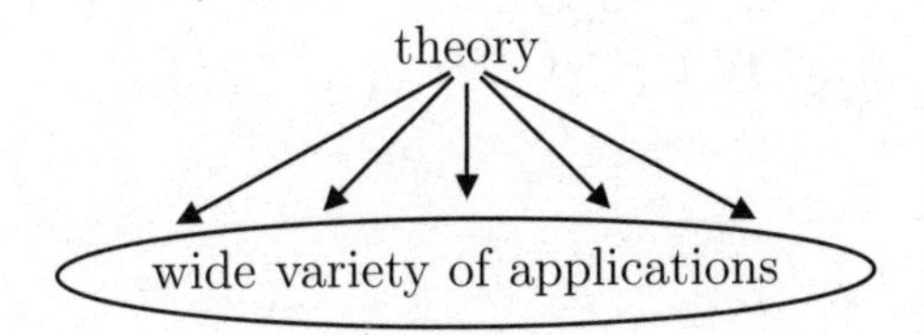

Something similar is true of the rectangular grid method for multiplying numbers: not only can we use the same method for any numbers, we can also generalize in some other ways. We can use it for three-digit numbers, just using a bigger grid.

	200	10	8
100	20,000	1,000	800
20	4,000	200	160
4	800	40	32

We now have to add these nine numbers together, which is a bit tedious, so perhaps at this point the scheme has become more illuminating than practical; then again, if we really want a practical and efficient method for multiplying three-digit numbers together I say we can just whip out a calculator on our phone.

We could try and use it for multiplying three numbers together, but at that point we need a three-dimensional diagram and it becomes somewhat harder to draw, so it's probably more of an abstract tool for thinking about things rather than for actually calculating things.

But we can also generalize it to multiply other things together, like letters. We could use this method to multiply $x + 2$ and $3x + 1$, or $(a + b)$ and $(c + d)$.

	x	2
$3x$	$3x^2$	$6x$
1	x	2

$$\text{total} = 3x^2 + 7x + 2$$

	c	d
a	ac	ad
b	bc	bd

$$\text{total} = ac + ad + bc + bd$$

This way of multiplying out parentheses is deeper and more illuminating to me than the dreaded mnemonic "FOIL" which stands for "First, Outer, Inner, Last" to remind us that we need to multiply the first things, the outer things, the inner things and the last things.

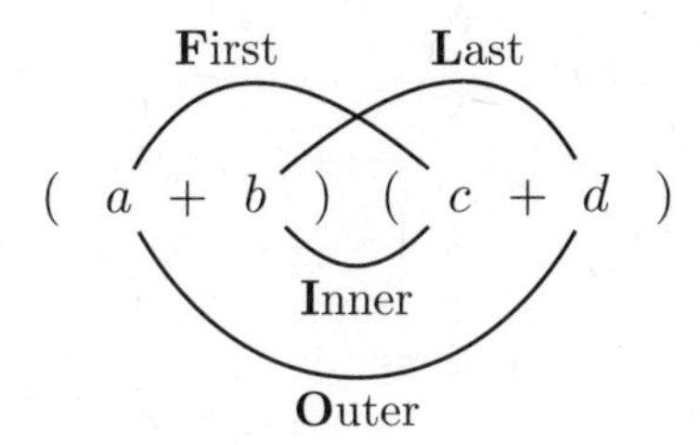

Mathematicians loathe this just as much as math-phobic students, or possibly even more; I'll come back to this in Chapter 6. Really FOIL is just an unilluminating version of the grid method.

So far we have used the grid method to multiply numbers and letters, showing a little of the method's satisfying potential for generalization. I'd now like to demonstrate a further generalization to a world of more complicated numbers: actually, the world of complex numbers.

Complex numbers

You might never have heard of complex numbers before, or you might have heard of them but forgotten what they are. Complex numbers are a type of number that's further down the line of "breaking rules" than the real numbers. We have this chain of successive rule breaking:

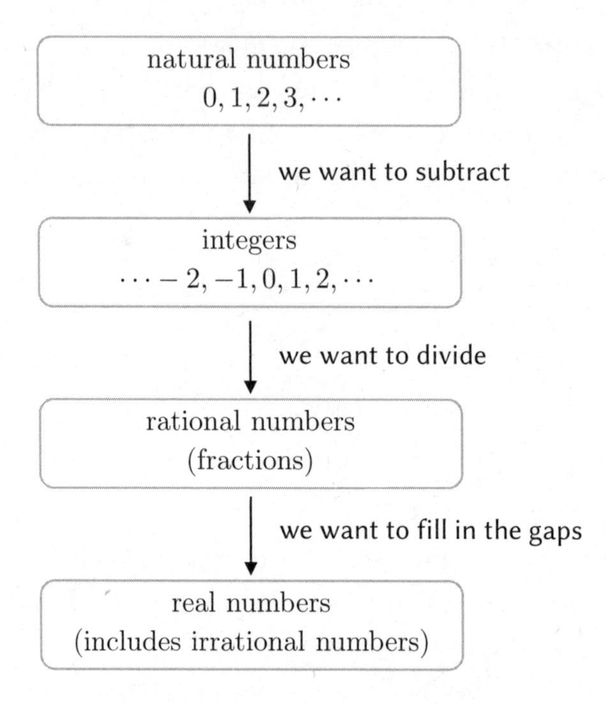

The next level of rule breaking happens when we feel frustrated that we can't take square roots of negative numbers. Maybe you've never felt that frustration. I myself don't remember that one specifically; I do remember feeling frustrated that we could only subtract smaller numbers from bigger ones, not bigger ones from smaller ones. I also remember feeling frustrated during an exercise in kindergarten where we drew around our "hand span," cut it out, and then went around the classroom measuring things. I was frustrated that nothing was a whole number of hand spans. We were allowed to say "two and a bit" but it seemed misleading to me that a "bit" could be tiny, or a "bit" could be almost an entire hand.

Anyway, later on in school you might have experienced being told that you're not allowed to take the square root of a negative number, and then the following year they announce "Now we're going to take the square root of a negative number."

This unfortunately makes it sound like mathematicians keep changing the rules, but the misleading part was really where anyone said "you can't take the square root of a negative number" in the first place. Perhaps you now wonder why that's true, and perhaps, given the sorts of answers I've explored so far, you realize that the answer comes down to definitions and contexts and that the real question is "*Where* can and can't we take the square root of a negative number?" In math, "Where can we do this?" means "In what world can we get a sensible answer to this?"

The most nuanced world we've explored so far is the world of ordinary (real) numbers, and in this world we really can't get an answer for the square root of a negative number. The reason comes from thinking about what a square root is. A square root of 4 is a number such that when you square it (multiply it by itself) the answer is 4. We know that 2×2 is 4, so 2 is a square root of 4. Also -2×-2 is 4, so -2 is another square root of 4.

Now here's the thing. If you multiply a positive number by itself it is positive. If you multiply a negative number by itself it's also positive. (And if you multiply 0 by itself it's 0.) This doesn't leave any options for a number that you can multiply by itself to give something negative—we've run out of choices. Here's a summary, using the letter x to represent the square root we're trying to find:

Can we find x such that x^2 is negative?

- If x is positive then x^2 is positive.
- If x is negative then x^2 is positive.
- If x is zero then x^2 is zero.

So x can't be positive, negative, or zero.

If x can't be positive, negative, or zero, this might seem to rule out all the possibilities. But it only rules out all the possibilities *in the world of positive and negative numbers and zero.* This rules out all the integers (whole numbers including negatives and zero); it also rules out all the rational numbers (which includes fractions) and all the real numbers (which includes irrational numbers). In all those worlds, all the numbers are positive, negative, or zero. What if there were some other world in which there are numbers that are not positive, negative, or zero? How is that possible?

Well what mathematicians do is just make something up. That's what we really did for negative numbers and fractions as well, it was just less obvious because we're more familiar with negative numbers and fractions. It was a bit more obvious with irrational numbers, because for those we had to invent a concept of "limit" for a sequence.

So in a fit of hilarity (or that's how I like to think of it), we invent an answer for the square root of −1. We pluck it out of our imagination, and because we did that, we call it an "imaginary number" and write it as i. It's a new building block. It's as if Lego just came up with a new type of special brick and the first thing we want to do is explore all the things we can now do if we add an unlimited supply of this new brick to our old stash of Lego.

As I mentioned in Chapter 2, some people think these shouldn't be called numbers. You're welcome to think that, but mathematicians have decided to call them imaginary numbers and that makes perfectly good sense, because they do behave a lot like numbers: we can do addition and multiplication, and implement an interaction between addition and multiplication, as we'll see.

Perhaps by "number" some other people mean "something that represents a length in the real world," but that is a closed and fixed characterization. Abstract mathematicians prefer characterizations by behavior rather than by intrinsic characteristic, and I argue that this is more open and inclusive, because it accepts any object that behaves in relevant ways. This is just like saying a mathematician is someone who can do mathematics, rather than saying that a mathematician has to be a man, and that you therefore won't allow women to study mathematics.

The point about calling these things numbers is that we can—and do—incorporate them in a world that behaves like numbers. The first thing to wonder is how to add them up. What do you think $i + i$ should be? It is quite sensible to call that $2i$. Even if we don't exactly know what i "is," we can consider having 2 of them. In fact we can consider b of them for any real number b, and we could call that bi. (You might want to call it $b \times i$ but mathematicians get bored of the × sign, so we write $2i$ instead of $2 \times i$ and we write bi instead of $b \times i$.) We can then think about adding different amounts of i together, and just like with apples, bananas or cookies, if we add $2i$ and $3i$ we gather them together and get $5i$.

Now what about if we add one of those "imaginary numbers" to a real number. For example what if we add 1 and i? Well that's like adding an apple and a banana. There's not a lot you can say about that except that now you have an apple and a banana. (A student of mine once suggested a smoothie.) So we have the answer $1 + i$. That might seem unsatisfactory, but we might also employ the slogan "it is what it is" (which I do find quite unsatisfactory in normal life). There really is nothing more we can say about it at this point: it just is 1 added to i. But then we have the possibility of all combinations like that, things like $1 + 2i$, $-4 + 3i$, and indeed $a + bi$ for any real numbers a and b. Things then do get more interesting, as we can depict these on a two-dimensional diagram like this:

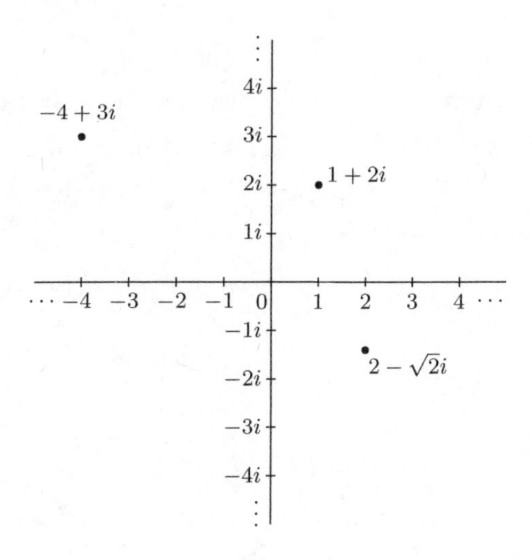

These are called *complex numbers*, because, well, they're sort of complex. Each one has a real part and an imaginary part (although either part could be zero), a bit like an x coordinate and a y coordinate, so they live in a two-dimensional plane rather than on a one-dimensional line. We draw the new "imaginary" direction at right angles to the real direction to show that it's completely different. Incidentally this doesn't mean the imaginary direction is unreal—"real" is a technical term here. There's definitely something more concrete about the so-called real numbers, because it's easier to say what they measure in the concrete world, although we've also seen that this already gets abstract if we're thinking about negative numbers.

Anyway so far we've multiplied real numbers by imaginary numbers, and we've added real numbers to imaginary numbers. How about adding and multiplying fully complex numbers, both of which have a real and an imaginary part? To add them up, it's like gathering apples and bananas again. For example, to add $(2 + 3i)$ and $(1 + 4i)$ we gather the real parts and do $2 + 1 = 3$, and we gather the imaginary parts and do $3i + 4i = 7i$ and the total is $3 + 7i$. We can also make use of column addition again and do it like this:

$$\begin{array}{r} 2 + 3i \\ 1 + 4i \\ \hline 3 + 7i \end{array}$$

For multiplication we can use the grid method again. What follows might be somewhat technical, so feel free to skim it if you're happy to accept the main point, which is that we can multiply these things using a method we've already seen.

For example if we want to try $(2 + 3i) \times (1 + 4i)$ we can draw this grid:

	2	$3i$
1	2	$3i$
$4i$	$8i$	$12i^2$

This works just like before; we just have to be a bit careful when we do the bottom right square, which is $3i \times 4i$. This gives $12i^2$, and then we need to know what i^2 is. Well, the entire point of i in the first place was for it to be the square root of -1, and this means that i^2 is -1, so:

$$\begin{aligned} 3i \times 4i &= 12i^2 \\ &= 12 \times (-1) \\ &= -12 \end{aligned}$$

So all in all we have this:

	2	$3i$
1	2	$3i$
$4i$	$8i$	-12

$$\begin{aligned} \text{total} &= (2 - 12) + (8i + 3i) \\ &= -10 + 11i \end{aligned}$$

Don't worry if you didn't follow all of that—I just wanted to show that the method of multiplying numbers in a grid was very widely applicable, and it managed to unify multiplication in many different contexts. This is something that mathematicians value in a piece of work even at a research level. If it solves a longstanding problem that's one thing, but often it has methods that help with other things too or, even better, it might have proceeded by finding a way to relate two different topics so that the understanding of each can be shared, unified, and built upon. This is a way of making progress in math: not just doing things we couldn't previously do, but understanding and building more and more complicated arguments and structures. This is something exemplified by complex numbers.

Building complexity

The word "complex" in the context of complex numbers means something precise and technical: it means that we are dealing with the world of real and imaginary numbers mixed together. But just as imaginary numbers are called that to evoke certain imagery, complex numbers are called that to evoke the imagery of general complexity. In a sense, all of math is imaginary, and all of it is complex. But also, in a sense, imaginary numbers are more imaginary than real ones, because they don't measure concrete things in the world. And complex numbers are more complex than real ones, because they mix up different structures (real and imaginary) to make a more complicated world. Complex numbers can be thought of as a higher-dimensional version of numbers, as they sit in a two-dimensional plane rather than a one-dimensional line.

It might seem that we were just building up complexity for the sake of it. But the wonderful thing is that although we did just sort of make up complex numbers from our imagination, it turns out that they enable us to build whole new branches of math, and understand things that we couldn't understand before, including significant swathes of physics. There's something about going into higher dimensions that gives us more insight into the low dimensions, even if it's only ever the lower dimensions that we're really trying to understand. Complex numbers do something like that. We start in the one-dimensional world of real numbers, and that's really all there is in our concrete world. But we use our imagination to include complex numbers, and then all sorts of calculations turn out to make more sense. There are patterns we can see that we couldn't see when we were all squashed into one dimension.

But where do those patterns exist, and are they real? (I mean that in the non-technical sense of "real.") In some sense it's true that they don't exist anywhere—nowhere concrete anyway. In another sense they exist in our heads, and who's to say that means they don't exist?

Those patterns in our heads help us understand real things about the real world around us.

To me this is like the issue of 0.9 repeating. In a sense, we just invented the answer to that, that is, we invented a way to make sense of adding up infinite sequences of numbers that are becoming infinitely small. But we didn't just pluck the answer to $0.\overline{9}$ out of our imagination, we came up with a rigorous framework for the *process* of reasoning with repeating decimals, and that process gave us an answer. The fact that we did it in a logically rigorous way means that we can build on it, and in that case it enabled us to build the entire field of calculus, on which almost every modern aspect of life depends. Then, putting calculus together with complex numbers gives the field of complex analysis, which is what a lot of modern physics depends on.

And this is really the point of logical rigor and abstract math. It's the fact that it enables us to build. It enables us to build complicated arguments knowing that the logic holds up. It enables us to package up complicated concepts and then treat them as building blocks so that we can understand increasingly complicated worlds of concepts, and create things that are further and further from our basic building blocks.

But this does bring us to an uncomfortable question: Why is that a good thing? In fact, is that a good thing? So far in this book we've been discussing what math is, how it works, why we do it, and what makes it good. But this has all been referring to a certain kind of math: rigorous formal math as defined by academic mathematicians, mostly European white men in the last few hundred years. Other mathematics led up to the definition of that framework, and other mathematicians have participated in the framework since it was developed, but the control and influence of those European white men is indelible. They came up with the framework in order to put mathematics on a secure footing, so that we can build stronger arguments and develop more complex systems. And the framework has been successful: mathematics has made extraordinary progress since then, as mathematicians have been able to reach broad consensus on what is true, and build on it. But is that necessarily a good thing? This brings us to questions of what we value more broadly in the world, and the question of what other types of mathematics have been overlooked, undervalued, or even suppressed in the process.

Progress and colonialism

Is progress necessarily a good thing?

There are some assumptions we're making here about the idea of "progress" and I'm going to argue, probably contentiously, that this notion of progress is not something we should assume is a good thing. It's a notion of progress that is wrapped up with the destruction of the earth's natural resources, and indeed with colonialism, and (largely white) colonialists' urges to wipe out more traditional cultures in the name of "civilization" and "progress."

There's a field of mathematics called ethnomathematics by mainstream academia. There are different definitions of it (as there are of all academic fields), but the idea is that this is mathematics that is more rooted in culture, rather than in the sterile academic process of logic and rigor. (I'm mixing my metaphors here because a sterile environment is a particularly good place in which to grow a culture in a lab.) And here "culture" means the culture of preexisting cultural groups, not the culture that has developed after the fact in academia.

The term "ethnomathematics" has an academic meaning, but it also risks an unfortunate connotation of "ethnic mathematics," which sounds like the mathematics of non-white people. This is because the term "ethnic" in normal life has unfortunately come to mean non-white to many people, because the dominant white culture tends to view white people as not having ethnicity, and view non-white people as "ethnic." And it is true that the topics of ethnomathematics typically come from non-white cultural groups.

Discussion of all these topics is extremely fraught and sensitive, with various groups of people very wary of attack from other groups of people. Unfortunately many groups have been excluded and attacked by mainstream mathematics and so this wariness is quite warranted. It does make these discussions extremely tricky.

First it's important to note that, historically, mathematics wasn't all done by white people. In fact all the great early developments in mathematics came from ancient non-white cultures: Mayan, Egyptian, Indian, Chinese, and Arab civilizations all played an important role in the development of mathematical ideas long before those ideas became dominated by white people. The ancient Greeks did a lot of thinking about math and philosophy, but mathematician Jonathan Farley pointed out to me that some of the people we refer to as "ancient Greeks" were not really Greek, but were from other parts of the Greek Empire, including parts of Africa. For example Eratosthenes, who came up with an ingenious way to find prime numbers, was from Cyrene (now part of modern-day Libya). Euclid, of geometry fame, was referred to at the time as Euclid of Alexandria, which might mean that he moved there as it was the center of learning at the time, but it might mean that he was really from there (in Egypt). It is imperialist of us to refer to everyone from the Greek Empire as "Greek." It would be like referring to Ramanujan as a British mathematician because he was from India when it was part of the British Empire—but now I'm jumping ahead of myself, and I'll come back to Ramanujan in a moment.

I am not at all an expert on these issues, but I believe it's important for us all to think about them. I have fallen into many traps when thinking about colonialism and mathematics, and I'm sure I'll keep falling into them, as will anyone who's been trained by a Eurocentric, white-dominated education system. However, I try to think about it and risk falling into those traps, rather than just stay safely in pure math research. And the fact that I'm not white doesn't make me immune from accusations of colonialism; at the same time, there are non-white people who think that worrying about the white-washing of mathematics is nonsense, but that doesn't mean it's nonsense.

Unfortunately oppression and exclusion mean there will always be oppressed people who feel (consciously or otherwise) that the safest course of action is to side with the oppressors, hence there are women who are anti-feminist, and there are Black people who support a political party that criminalizes Black people and suppresses Black people's votes, and there are Asian immigrants who support restrictions on immigration. That doesn't make any of those oppressions valid, it just exemplifies how strong the forces of oppression are.

In all, there is rather a lot of nuance that we need to consider. First of all, we need to acknowledge that all the early developments in mathematics were by non-white people. Then, we need to acknowledge that contemporary mathematics has been exceedingly dominated by white people, has been largely gate-kept by white people, and also overrepresents white people. Most (but indeed not all) modern developments in mathematics are credited to white people. Pointing this out does not in itself constitute erasing the non-white history of mathematics.

Now, at the same time as all this, it is true that mathematics as a field has grown carefully and rigorously according to a carefully constructed framework of logic and rigor. It is, arguably, the *framework* that is keeping some math in and keeping some math out. But even aside from the huge question of inequitable access to resources and education, there are questions about the values built into the framework of mathematics. That framework is geared toward the principles of "progress" and "development." Deep down I have an uncomfortable suspicion that those principles are indelibly linked to colonialism, imperialism, and the urge to conquer others.

It's true that non-white cultures have participated in the conquering of other people, to greater and lesser extents. Also white people have tried to conquer other white people, not just non-white people. But I do think that the current world order is inextricably tied up with the overwhelming way in which white people have overpowered non-white people with their "developments," if nothing else, their ability to develop more and more destructive weapons and other machines of war, which has historically always been the prime way of conquering less "developed" nations. In the twenty-first century there are other, less directly destructive perhaps, but more insidiously destructive weapons, via technology and the way that capitalism concentrates power among the already powerful: at the level of individuals, and also at the level of countries.

The very way in which we talk about "developed" countries and "developing" countries has all of our judgment about development built into it.

But let me take a step back for a second and acknowledge that at some level I, personally, have bought into this system of values. I love mathematics because I love its ability to build, develop, and make progress, which depends on the strong framework it has for assessing truth and building consensus. There are many other things I love for similar reasons. I love Western classical music† the most, out of all kinds of music, and my favorite music is the kind that has the most development in it, the most complex structures, the kind of music that could never be improvised on the fly or passed down from generation to generation aurally, because of its sheer structural complexity. I'm not claiming that other music isn't complex; what I'm trying to describe is a very deliberate form of complexity where things are written down and consciously constructed, structure upon structure. I like complex, structured literature that has threads winding together in a tight pattern, the sort of literature that could not possibly have been typed in one go on a single roll of paper, but must have been meticulously planned to make sure all the pieces fit together. I love food that involves structure and development, sauces that involve a careful

† This is widely known in Eurocentric cultures as just "classical music," so calling it "Western classical music" is a standard way to acknowledge that other cultures also have classical music. However, the term "Western" is itself problematic, and doesn't really make sense. The "Far East" is not very far east if you're in it.

technique to transform a combination of ingredients into something magically different from what we started with.

All of this involves development, and you could argue that my love of development is simply an aesthetic. But I am self-critical, and worry about things like structural exploitation and colonialism, so I worry that the love of development is tied to colonialism and to the imperialist view that a developed nation is *superior* to a developing one, and that this is why some countries are richer than others, and why some cultures have overpowered others.

And inside that question I feel called to ask myself: Does this incessant development mean we're better? Does it make us better in any framework other than the one we've constructed?

There are cultures that do things differently, and who are we to judge that our way is better? One of the most vivid examples of this cultural difference is in the story of Ramanujan.

Ramanujan and Hardy

Srinivasa Ramanujan was a brilliant mathematician from India, with an interesting and tragic story. He was born in India, in 1887, and had no formal mathematical training. He did win a scholarship to the Government Arts College, Kumbakonam, but he kept doing his own thing instead of following the formal demands of the British-style curriculum, which resulted in him failing most subjects except mathematics, and losing the scholarship. He later attempted further education again, this time at the non-British funded Pachaiyappa's College, but he still didn't fit into the required mold, and failed to get a degree.

He lived in extreme poverty, scraping together a living while pursuing mathematics in his own time and being convinced that the deep truths of mathematics were given to him by a goddess.

In 1913, Ramanujan wrote a letter to the "conventional" mathematician Professor G. H. Hardy at the University of Cambridge—

conventional, at least, in the framework of contemporary, European style mathematics. Hardy had done all the things the framework said he should: he had done a formal degree in mathematics, written a thesis, written meticulous proofs of things and published them in peer-reviewed research journals.

Ramanujan had done none of those things, but Hardy recognized his brilliance and brought him to Cambridge to study formal mathematics in the contemporary European style. This was a very big step for Ramanujan as his religion dictated that he should not go overseas, and his mother was very averse to his going.

When Ramanujan arrived, there was a direct clash of cultures at all levels. Hardy insisted that Ramanujan learn how to prove results according to the frameworks of European mathematics, to make sure the results were really true. Ramanujan did not see the point of doing that, when the goddess had already spoken those truths to him. Eventually Ramanujan was persuaded, partly by Hardy pointing out to him a flaw in one of the things he thought was true.

Incidentally, while it is true that Ramanujan believed he was being spoken to by a goddess, his innate and apparently magical affinity for numbers is sometimes rather overstated by people wanting to romanticize his "genius." There is the much-told story that when Ramanujan was ill in hospital, Hardy visited him and commented that his cab number was 1729, which he described as "not a very interesting number." Ramanujan immediately pointed out that, on the contrary, this is a very interesting number as it is the smallest number that can be expressed as the sum of two cubes in two different ways:

$$\begin{aligned} 1729 &= 1^3 + 12^3 \\ &= 9^3 + 10^3 \end{aligned}$$

This immediate riposte has often been taken as a sign of Ramanujan's genius, and his ability to immediately see things in numbers that even the great number theorist Hardy could not.

However, close examination of his notebooks later revealed that he had been studying "near misses" to Fermat's Last Theorem. Fermat's Last Theorem is the famous thing that the mathematician Pierre de Fermat scrawled in a margin of a book in around 1637, along with the comment "I have discovered a truly marvelous proof of this, which this margin is too narrow to contain." The theorem is that there are no solutions, in whole numbers, to the equation

$$x^n + y^n = z^n$$

if n is 3 or more. For $n = 2$ we know there are solutions and this is related to Pythagoras's theorem about the lengths of sides of a right-angled triangle. You might have been told to learn, in school, that two handy examples of right-angled triangles are 3, 4, 5, and 5, 12, 13. (That's the lengths of the three sides.) I remember, when I was at school, test writers were obsessed with everything boiling down to those two triangles. I suspect it's because the people who write exams were from a generation before calculators, when it was hard to find solutions to Pythagoras as it involved finding a square root. Finding square roots in general, without a calculator, is hard.

It's actually not really fair to call it "Fermat's Last Theorem" because he died without ever telling anyone what his amazing proof was, and usually until you've proved something it doesn't count as a theorem. It was not proved until 1994, when Andrew Wiles proved it; in fact his first proof from a year earlier had turned out to have a mistake but he then succeeded in correcting it. In any case the eventual proof depended on mathematical advances far beyond the time of Fermat, so there is no way this can have been Fermat's idea for a proof. Mathematicians currently believe that Fermat was mistaken about his proof, and they even have an idea of exactly how he might have been mistaken.

Anyway, many decades earlier, Ramanujan was investigating "near misses" to Fermat's Last Theorem, and among them numbers that are 1 away from being solutions in the following sense: instead of

$$x^3 + y^3 = z^3$$

a "near miss" is something like

$$x^3 + y^3 = z^3 + 1$$

He had already come up with 1729 as an example (as it is $12^3 + 1$), so when Hardy showed up with that taxi number it simply happened to be a number Ramanujan had already been studying, not something he plucked out of thin air. There is enough that really is mysterious and wonderful about his mind that we don't need to invent things.

What makes a great mathematician? The taxi number story, and Ramanujan's story in general, risk perpetuating the idea that to be a great mathematician you have to be a spectacular and mysterious genius, with an inexplicable and almost magical affinity for numbers and the ability to pluck truths from thin air.

I'm not sure what it takes to be a *great* anything, but to be a *good* mathematician you don't have to be any of those particular things. You need to be open-minded and be able to think flexibly, and be able to see things from many different points of view at the same time. You need to be able to see connections, which often means being able to ignore certain details about a situation in order to see how it matches up with another when those certain details are ignored. But you also need to be flexible enough to put those details back in, and ignore different ones to see things differently. You need to be able to construct highly rigorous arguments, hold them in your brain, move them around, and fit them together with other highly rigorous arguments. And you need a tolerance, or even a thirst, for the increase in manufactured complexity that this brings with it. This also involves creating ways to deal with that complexity, like creating special eggs and then creating a special egg carton to carry them around in. And then creating a special crate for the special egg cartons, and then perhaps a special truck for those special crates, and so on. Thus it often involves building up gradually bigger and bigger dreams from smaller ones, so it calls for a vivid imagination and ability to bring weird and wonderful ideas to life in your head. There is a myth that math and science are separate from "creative" subjects in the arts, but the line between them is really quite blurry. The myth probably comes from thinking that math is just about step-by-step computations with clear answers. But note that in describing a good mathematician, I did not at any point mention arithmetic, computation, memorization, numbers, or getting the right answers. Some computational parts of math do involve computation, but not all math is computational.

On the other hand I did mention a lot of building up of rigorous arguments. And that's where I betrayed the fact that I'm describing a good mathematician from the specific point of view of Western/European/colonized mathematics.[†]

Hardy judged Ramanujan from this specific point of view, found him to have deep promise but to be lacking technically, and urged him to meet those Western colonial standards. The question that taxes me is whether Hardy was right to insist that Ramanujan prove everything in the manner of European mathematics. He was right according to the framework of European mathematics, but that is a circular argument.

This is where another aspect of colonialism comes in. Ramanujan was not culturally accepted in Cambridge and became very ill because he was unable to find nourishment suitable to the requirements of his own beliefs. In particular he was a deeply religious Hindu and strict vegetarian, and that diet was more or less impossible in a Cambridge college at the time. (It wasn't even that great in my time, but it was at least possible.)

† There's the problematic term "Western" again.

Eventually he went back to India but never recovered, and died at the age of just thirty-two. Despite his youth, he was recognized by the reigning powers of British mathematics and elected a Fellow of the Royal Society, one of the youngest ever, and the second-ever Indian after Ardaseer Cursetjee in 1841. He was the first-ever Indian to be elected a Fellow of Trinity College (though they only did that *after* he was elected to the Royal Society).

This might sound like the triumph of a poor boy from India being accepted into the lofty circles of Cambridge mathematics, but seen from another light it's a stodgy institution insisting that someone from a different culture conform to their norms in order to be accepted.

When Ramanujan died he left behind him notebooks containing copious quantities of notes of further "truths" that he had not got around to proving in the European style. Mathematicians spent a hundred years working on them and to date almost all of it has been proved to be correct, according to European methodology. But Ramanujan was already convinced, for his own reasons.

Who is to say which method is better? Some of Ramanujan's results might have been slightly incorrect, but they still contained great insights, and the errors were illuminating ones. Indeed, Andrew Wiles's first attempt at a proof of Fermat's Last Theorem also turned out to be incorrect, so it's not like European mathematicians are immune from making mistakes either. There are numerous times that mathematicians have found errors in their own or other people's proofs, and papers have to be corrected or withdrawn.

In a way this reminds me of the battle between old-school publishing via gatekeepers and peer review, and newer methods of crowdsourcing such as Wikipedia. The old-school gatekeepers are often horrified by Wikipedia, because they think that the fact that *anyone* can contribute means it will be full of errors. And it does contain errors, but so do the peer-reviewed, gate-kept publications. *Nature* famously compared Wikipedia with *Encyclopaedia Britannica* in 2005, and Wikipedia didn't do badly at all.† That was still early days for Wikipedia, and they did even better in a comparison in 2012.‡ As it happens, that is also the year *Encyclopaedia Britannica* went out of print.

The difference with white-people's math is that on the very grand scheme of things it's not the old-fashioned people resisting change, it's the relative newcomers (European culture being relatively young) declaring the long-standing methods of ancient civilizations to be inferior. Why have some modern cultures come along and declared the old methods to be invalid? Contemporary culture is still baffled by how ancient cultures were able to do things like build Stonehenge, or construct the Pyramids. Perhaps there is more to the unproved, un-peer-reviewed methods than contemporary academics want to admit.

And this is the crux of the comparison between mathematics and ethnomathematics, or between mathematics based on development, which takes us as far away from natural culture as possible, and mathematics based in, and always remaining connected to, culture. We might marvel at, say, Inuit building a kayak without doing anything we would recognize as calculations, or similarly Amish people raising a barn—rather, they use a method that has been passed down from generation to generation as part of their culture. We might call it mathematics, to honor its brilliance; but is that imposing our own cultural norms on it? We might call it ethnomathematics to recognize that it is mathematics but in a slightly different vein from our proof-based one. But then, are we "othering" the math of other people?

---

† Jim Giles, "Internet Encyclopaedias Go Head to Head," *Nature* 438 (2005): 900–901, www.nature.com/articles/438900a.

‡ I. Casebourne, C. Davies, M. Fernandes, and N. Norman, "Assessing the Accuracy and Quality of Wikipedia Entries Compared to Popular Online Encyclopaedias: A Comparative Preliminary Study Across Disciplines in English, Spanish and Arabic" (Brighton, UK: Epic 2012), https://upload.wikimedia.org/wikipedia/commons/2/29/EPIC_Oxford_report.pdf.

And in the end, is our math better? Our relentless development and progress have led to the destruction of the environment we depend on for survival: the earth. Whereas native cultures, without the colonial/European/imperial inexorable drive for "progress," understand how to work in harmony with the surroundings, and be nourished by the environment without exhausting and destroying it.

Which of those things is the greater achievement, really: living in sustainable harmony with our environment, or pushing the development of industrial processes so dramatic and devastating that we now have to develop emergency processes to restore our destroyed environment?

If the latter is "progress," do we want it?

I don't have the answers, I just think it's important for us to take these questions seriously, and keep trying to do better.

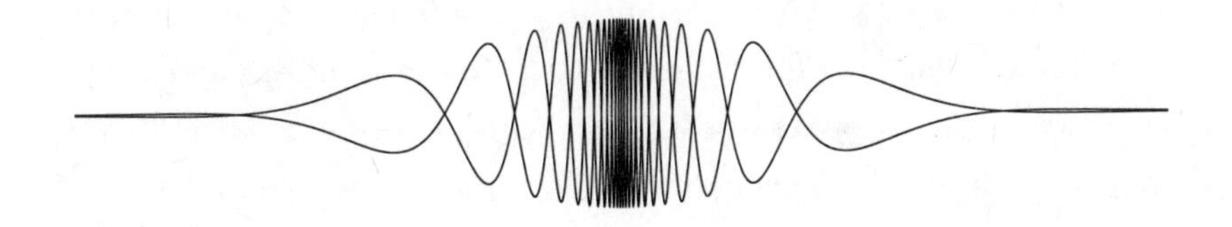

CHAPTER 5

LETTERS

Why does $y = mx + b$? In fact, what are those letters doing there at all?

In the first few chapters of the book we've been thinking about the general idea of math, where it comes from, how it works, why we do it, and what the point is. I'm now going to talk about more specific topics in math, and see how they often grow from innocent questions that need deep math to answer them. We'll begin, in this chapter, with the thorny question of using letters in math, which leads us into the topic of algebra.

Why do we turn numbers into letters? People often say to me, with a shudder, "I was fine with math, until the numbers *turned into letters* . . ."

So before even addressing $y = mx + b$ I'd like to address the question of why we turn numbers into letters in the first place. After we've discussed the motivation for doing that at all, we'll look at the reasons for doing it in the specific case of $y = mx + b$. Then finally we'll look at what this equation is telling us, and when it is and isn't true—because it's only true in particular contexts, even though it might come across as an absolute truth that you just have to remember.

I want to talk about it in that order because it's important to motivate math before doing it. When it comes to using letters instead of numbers, we've been hinting at it and been tugged toward it in the earlier part of the book, and this hinting and tugging is key all across math: the math we do pushes us toward the new math we come up with. It's just that if you don't feel that tug, or aren't shown that tug, then it feels like a contrived push rather than a natural flow. It's as if we're just throwing ingredients in a bowl to see what will happen, rather than having some reasoning behind our choice, even if it's mostly intuition or hard to explain. The fact that it's hard to explain makes it even more important to attempt to explain it.

To me this is all a bit like stepping on a paternoster. It's possible you've never stepped on a paternoster, the extraordinary type of elevator that never stops moving. It has a string of "capsules" without doors, that move in a big circle (or rather, a big, tall oval) through all the floors of a building. It's supposedly reminiscent how some Catholics use a string of beads to keep track of where they are when they're saying many repetitions of a prayer, which is, I believe, why the elevator version is called a paternoster. Anyway, the capsules never stop moving, and at any moment there's one for every floor going up, and one going down as well. So when you're standing on any particular floor there are two sides, the up direction and the down direction, and you just wait for the next open capsule to pass by and you step into it as it's moving.

There's one at the University of Sheffield, and people often go and ride on it just for fun, for the experience. I also had to take it to get to a regular meeting one semester. The first time I stepped on it I was utterly terrified. Then I found that getting off was even scarier than getting on. In both cases the key turned out to be anticipation, but it was very counterintuitive to anticipate the movement and allow myself to be naturally carried along with it. Rather than trying to step into the lift, it was better to stick my foot out and let the lift pick me up as it passed by. Getting out again was much harder; by symmetry, on the way down it was easier to get out than in.

The analogy I'm trying to make is that if you somehow feel you're being pushed into another level of math abstraction then it feels like a leap, and you might fall over trying to do it. I did once get my foot stuck in the paternoster. Luckily there is a safety mechanism in the form of a flap on the step that flips up, so I didn't really get stuck. Whereas if you feel the momentum of the math pick you up and carry you along it feels much more organic. Then, rather than thinking "Ugh, why are we doing letters?" you think "Phew, thank goodness we have a better way of saying this."

A better way of saying what?

There have already been some moments earlier on where I gave examples of things and then vaguely expected you to get the point, extrapolating from my example. Or where I had to use some very wordy phrasing in order to get the point across. For example, if we want to talk about adding things together, we might say that if we take one thing and five things we get six things, no matter whether we're talking about apples or cookies or bananas or anything else, as long as they don't spontaneously combust, combine, or reproduce. (And we don't eat them.) That is much more succinctly expressed as:

$$1x + 5x = 6x$$

The point isn't just succinctness for the sake of saving space, it's about packaging things up so that we can carry them around better. I always like to imagine those vacuum bags for clothes where you put your clothes in the bag and then attach a vacuum cleaner and suck the air out, so that the whole thing becomes very compressed and easier to carry around or store in a small space. Having a succinct way to express a situation involving a lot of possibilities means that we've compressed it down to one thing. The above expression using x's contains inside it the fact that one apple plus five apples is six apples, and one cookie plus five cookies is six cookies, and one elephant plus five elephants is six elephants, and so on. It has turned an infinite number of expressions into one expression.

This happened a lot when we were talking about the basic things we could do with numbers, in Chapter 3. At the time, I just gave some specific examples. I started with this:

> It doesn't matter what order we add things together, like
> $$2+5 = 5+2.$$

This gives the general idea, but it does leave you guessing the fact that this also means that 3 + 4 = 4 + 3 and 5 + 2 = 2 + 5 and so on. Whereas I can precisely express all of those possibilities by saying it like this:

> For any numbers a and b, $a+b = b+a.$

Similarly for how we group things together, I said

> It doesn't matter how we group things, like
> $$(2+5)+5 = 2+(5+5).$$

This gives one example of a general principle, which is not a complete picture. The complete statement of the general principle is this:

> For any numbers a, b, and c, $(a+b)+c = a+(b+c).$

We also saw a general grid pattern for adding odd and even numbers, and thinking about tolerance. We could describe it at some length in words, or we could express it succinctly using letters like this:

A	B
B	A

It might seem pedantic to worry about whether or not we've rigorously expressed the general idea, when the general idea seems pretty obvious from the specific example(s). And in many cases it might well be a little pedantic. But as I have said before, I consider pedantry to be precision without illumination, and so if there is some illumination to be had, then it no longer counts as pedantry, in my opinion. And I really do believe in illumination. I am not a grammar pedant, as I believe in communication more than in dogmatic adherence to grammatical rules, and if a grammatically correct construction is also going to sound pompous and alienating then I'd rather avoid it. (I do enjoy the contortions that would sometimes be required to avoid ending a sentence in a preposition—but I enjoy observing them without feeling compelled to implement them.) Likewise I dislike rules about commas: I am neither a stickler for the serial comma nor totally against it, I'm just against dogmatic rules for it, because I prefer to use each comma as seems fit for that sentence.† And yes there are sentences whose meaning completely changes with the addition or omission of the serial comma, but they're usually rather contrived, like the contrived situations in which it might be really necessary in life to be able to do mental arithmetic.

In the scenarios with numbers that I just described, it's possible that the situation was unambiguous enough that expressing it in letters didn't add that much. But as things get more complex it might become less obvious what's going on if we just give an example, or it might be seriously ambiguous. Standardized tests seem to love questions saying things like "What is the next number in this sequence?" These questions always wind me up, because all they ever do is give a few numbers and then expect you to decide what the next one "has" to be according to some unstated pattern. This should not be called mathematics, it should be called Psychic Powers, as you're supposed to work out by your psychic powers what the test writer was thinking.

† This bothers copy editors. Sorry to any copy editor who is helping me by reading this.

Logically, anything could come next. For example if the sequence goes like this: 2, 4, . . . we really can't say whether the next number is 6 or 8, or something else entirely. Even if we give more numbers at the beginning, say 2, 4, 6, 8, 10, then it might well be true that the most *obvious* next number would be 12, but maybe the sequence is the number of push-ups you're going to aim to do each day, and after five days you have a rest day, and then you start again but with an increased number. So that sequence would go like this:

$$2, 4, 6, 8, 10, 0, 3, 5, 7, 9, 11, 0, 4, 6, 8, 10, 12, 0, \cdots$$

Or maybe it's a sequence of numbers that goes like this:

$$2, 4, 6, 8, 10, 800, 7532, 15, \pi, -100000000000, \cdots$$

for no particular reason except to make the point that sequences of numbers can do absolutely anything, and if we only list a finite number of things in the sequence there is no logical way to be certain what comes next.

The point of math is to pin things down using logic, not guesswork or psychic powers, and this is one reason that writing things down according to letters comes in. It enables us to express general relationships between numbers, which apply to all numbers, rather than specific relationships that only apply to specific numbers.

Relationships

Math is really more about relationships than it sometimes seems. My field of research, category theory, makes that very precise and really tries to study things in terms of their relationships with other things, rather than by any of their intrinsic characteristics.

Even when we write down an equation, what we're really doing is expressing a relationship between some things. $1 + 1 = 2$ is a relationship between the numbers 1 and 2, and this is a more subtle way of thinking about it than thinking about it as a "fact."

But that's a relationship between some *specific* numbers, 1 and 2. If we want to express *general* relationships between numbers, we're looking to express relationships between any numbers, not just specific ones. For example, no matter what numbers we're thinking about, adding them together in different orders gives the same result. That is, for all numbers a and b,

$$a + b = b + a$$

This is a general relationship between numbers, not between specific examples of numbers. The more complex the general relationships involved, the more beneficial it becomes to express them with letters rather than trying to hint at them using numbers. Math is about spotting patterns, but if we can pin down the pattern precisely then that's more productive than leaving it to everyone's guesswork. We can then, in effect, write down an infinite number of relationships in one go, a bit like a child who draws an infinite number of lines of symmetry on a circle by coloring in the circle.

Suppose we want to tell someone we're thinking about an infinite sequence that starts like this:

$$0, 2, 4, 6, 8, 10, \cdots$$

and keeps going like that. It might be a little clearer if you know about odd and even numbers, and can say "skip each odd number and include the next even number." But mathematicians still find that a little unsatisfying, as it's still quite wordy, and leaves a feeling of walking through the sequence one step at a time. Mathematicians like to express an infinite sequence by saying what the nth term is for each natural number[†] n. So for our sequence of even numbers we could say something like this:

For any natural number n, the nth term in the sequence is $2n$.

We can make this even more succinct by calling the nth term in the sequence a_n, and then we just say

For any natural number n, $a_n = 2n$.

Now it's no longer ambiguous what's happening in the entire sequence, as we've completely pinned it down according to logic.[‡]

I know that those are still extremely contrived examples, but this idea of representing general quantities by letters is a powerful idea that opens up many possibilities in math. This is what is thought of as "algebra" in school math, although that's quite different from the concept of algebra in research math. The word algebra comes from the Arabic *al-jabr*, which means something like "re-union of broken parts," and was originally used to describe the setting of broken bones. Its mathematical use was coined by the ninth-century Persian mathematician al-Khwarizmi, who was writing about manipulating symbols in equations (as in school algebra), but research algebra is really about putting parts together.

Here's an aside on how we give credit to mathematicians who were the first to do something. It is definitely "correct" practice to give credit to the first person who came up with a mathematical idea, and

† Remember, the natural numbers are the counting numbers. Here I'm taking them to be $0, 1, 2, 3, \cdots$ and so on.

‡ In case you're wondering how to express the sequence of push-ups, I admit it would be rather complicated. We'd have to say something like: every natural number n is of the form $6k + r$ where k and r are integers and $0 \leq r < 6$. Then $a_{6k+r} = k + 2r$ if $r > 0$ and 0 if $r = 0$. We should also say that this sequence starts with $n = 1$.

it is polite to do so. It would certainly be wrong to claim that the originator of the idea was someone else. However, I personally think that mathematics stands or falls on its own logic, not on the reputations of the people who came up with it, and so I don't think it's as important as it is in most (perhaps all) other subjects to draw attention to the people who did things. On the reverse side, I believe it can actually be a bad thing to draw too much attention to the humans who came up with the ideas, because math really is not supposed to depend on humans, but on the framework of logic. Moreover, sometimes many mathematicians came up with the same ideas at around the same time, and our obsession with who did things first is not quite the point.

However, there's another consideration when it comes to drawing attention to the fact that some important progress was made by non-white people, and this is to do with fighting white supremacy in mathematics. Unfortunately there are still people who think of math as a white male domain, and the more examples we have to combat that, the better. In the last few hundred years white men have *made* math into a white male domain, but they did it by exclusion, not by any sort of innate ability, and we can and should correct for that injustice.

That's the end of the aside. I now want to talk about what we can go on to do, once we've started expressing things using letters. One thing we can now do is build the complexity, so that we can combine relationships into more complex ones. This is the point of substitution, where we combine relations. So we might know that $a = b^2$ and $b = c + 1$ and then we can combine those to discover that $a = (c + 1)^2$. That's not an interesting example at all (like so many examples in math texts, whoops), but we do pile up relationships in life as well, for example when we talk about aunts and uncles, which most people are fairly comfortable understanding, and then move on to cousins or second cousins once removed, which many people find very confusing.

Using letters can help us get further with that sort of "piling up," but it can be very exhausting on the brain if you're uncomfortable with the very fact of having used letters in the first place. It's a bit like the fact that you can travel much further on a bicycle than on foot, unless you have trouble riding a bicycle.

It's true that letters are more abstract than numbers. But numbers were already more abstract than the things they were unifying, and most of us have managed to get our heads around that abstraction, more or less—we even did it when we were really quite young. This shows we're all capable of it, it can just be baffling if you don't see why you're doing it, in which case you have no motivation to do it. There are plenty of things I'm pretty sure I could learn how to do if I had more motivation (such as catching a ball, or ironing), but I have no motivation to do it, so I am still not able to do it. Not very well, anyway. And I admit that at a certain point in my life it became more comfortable to refuse to value it as a skill, than to continue to put up with being ridiculed for being so bad at it. I see this coming into play with people loudly declaring they're no good at math and insisting that it's pointless anyway. The remedy in that case isn't to stress how useful it is, but to stop belittling people who find it difficult.

Now, it's true that motivation only gets us so far. There are other things I really can't do no matter how motivated I am, such as teleportation. I am highly motivated to do that, but alas still can't do it. So motivation can't necessarily do things for us, but lack of motivation can certainly get in the way.

It might seem that abstraction is a niche activity, but we are motivated to use abstraction in some commonplace situations without even noticing it. Pronouns are also an abstraction that can be confusing and cause mental overload. Pronouns enable us to refer to someone without repeating their name over and over again, but they also enable us to refer to unspecific people, general people rather than specific people, as I just did in that sentence. I said

> Pronouns enable us to refer to *someone* without repeating *their* name.

This sentence applies to any people. If I had to use examples of specific people, I'd have to say something like "Pronouns enable us to refer to Emily without repeating Emily's name, and to Tom without repeating Tom's name, and to Steve without repeating Steve's name, and so on," which would be much more tedious. Using letters to refer to numbers is doing the same thing in math. We sometimes do that with people, too, when a situation has become too complicated to keep track of it by pronouns alone, and we might talk about a situation in which person A does something to person B and person B responds by doing something to person C, who then does something to person A.

Incidentally, the third-person gender-neutral singular pronoun "they" is another level more abstract, as it enables us to refer to someone without knowing their name or their gender. Some of us use it when we don't want to specify gender, either because we know that someone is nonbinary, or because we don't know what someone's gender is and don't want to perpetuate the implicit bias of generally using "he" as was standard in the last century, and perhaps we find "he or she" unnecessarily cumbersome as well as erasing nonbinary people. This use of "they" is a level of abstraction too far for some people, but I'm sure everyone is capable of doing it, if they feel motivated enough. The problem is that some people don't feel motivated to do it, because they don't see the problem with gendered pronouns; even worse, some people specifically decline to be inclusive toward nonbinary people.

With letters the point is the one about not knowing: we want to be able to refer to something when we don't know what it is yet. Sometimes the point is to write down relationships between quantities when we don't know what those quantities are yet, and then use the relationships to *deduce* what they are. Unfortunately yet again the examples are often extremely contrived, like this old chestnut:

> My mother is three times my age but in ten years she will be twice my age. How old am I?

Math is like a detective story. We have clues, we piece them together, and we deduce things we didn't know before. And it's much easier to piece the clues together if we have a way of referring to things before we know what they are. This method doesn't just make deduction processes easier, it actually enables us to have thoughts and ideas that would otherwise almost certainly have been impossible. It's still helpful to try the method out on some more straightforward situations though, like straight lines. This is where the equation $y = mx + b$ comes in: it's an equation describing straight lines in a particular form of two-dimensional space. How do we know it's right? Well first of all we need to know what any of this means, starting with how we describe *points* in two-dimensional space, let alone lines.

Two-dimensional space

The x and y in the equation are referring to x and y coordinates, in a two-dimensional plane. Already we've made a choice—representing two-dimensional space in this way is just one possible way of describing the world around us. It's called "cartesian" because the scheme was first published by the mathematician and philosopher René Descartes, in the seventeenth century. Here's a picture of two-dimensional space organized according to x coordinates and y coordinates:

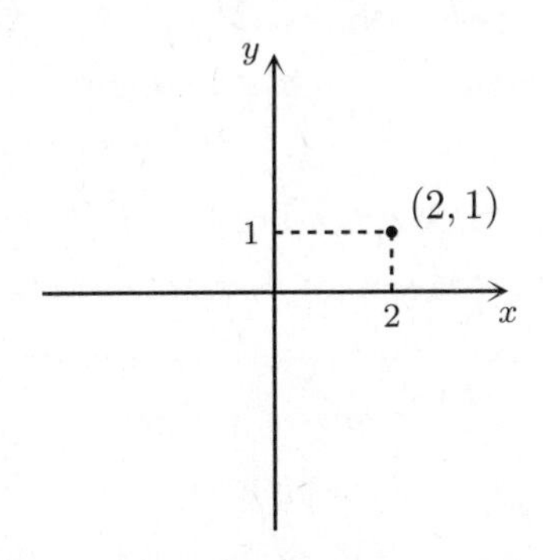

The idea is that any point in the plane can be unambiguously described by giving its two coordinates, representing how far along the point is in the x direction, and how far along it is in the y direction. By convention, we usually give the x coordinate first, so the point I've marked in that diagram is at the position (2, 1).

However, here is another perfectly rigorous way to pin down two directions, using a skew grid:

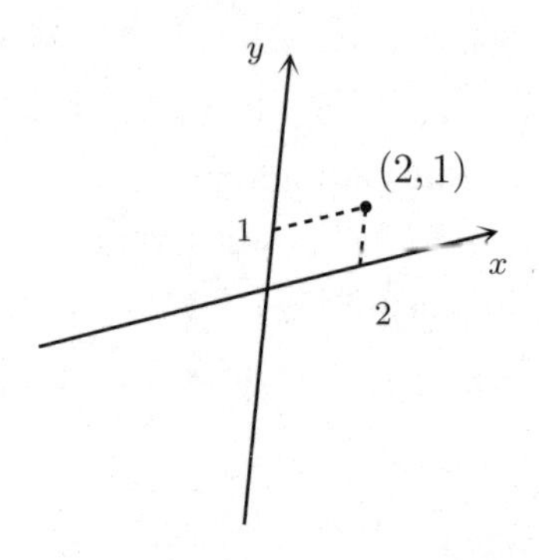

There's no particular reason to use axes at right angles, except that it's slightly more convenient in some ways, and perhaps more intuitive. Abstractly a pair of axes at another angle is just as good, and mathematicians do study the relationships between how we express things according to one choice of axes, and how we express things relative to a different choice. It's useful to be able to switch your frame of reference, but only if you understand how everything you're referring to transforms in that process, otherwise you won't be able to recognize when you've got the same thing in a different context, as opposed to a genuinely different thing.

Also note that there are ways to think about two-dimensional space that don't even use two axes at all. If you look at a map of your country, the lines of longitude and latitude will look more or less like a cartesian grid. However, if you zoom in on a map around the North or South Pole, you'll see that the lines of longitude and latitude don't look like a cartesian grid anymore: the lines of latitude look like concentric circles, and the lines of longitude look like the spokes of a bicycle wheel.

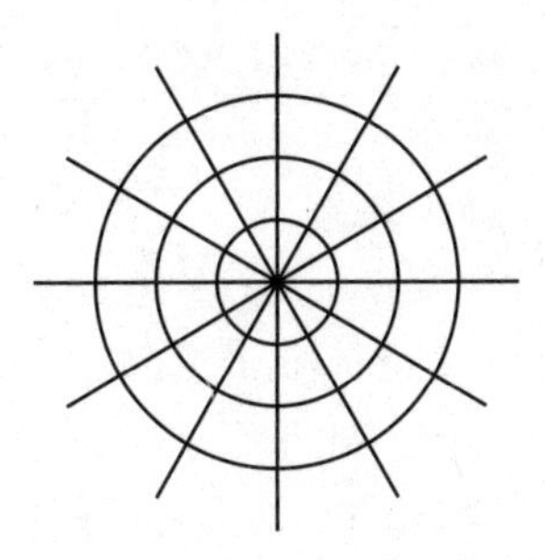

Even if we're not at a Pole, we could use this scheme to pin down a position in two-dimensional space any time we want, if we feel like it. Instead of an x coordinate and a y coordinate, we would specify the distance from a chosen center point, and the angle from the horizontal (say). Those coordinates are called polar coordinates because it looks like what happens at the North and South Pole. Note that there's nothing extremely different about the shape of the earth at the Poles, it's just that the way we have chosen to place the coordinates of longitude and latitude means that those are the places where this circular scheme appears.

There are some situations in which it makes particular sense to use the circular form of coordinates instead of cartesian ones, for example if you're using radar to sweep out a zone around a watchtower. In that case the sensor moves in circles, and at any given moment it senses the distance of a detected object from the central tower. This means that the two pieces of information we most naturally have are the angle of the sensor, and the distance, in a straight line, from the object to the sensor. Here's a schematic diagram of the situation as viewed from above:

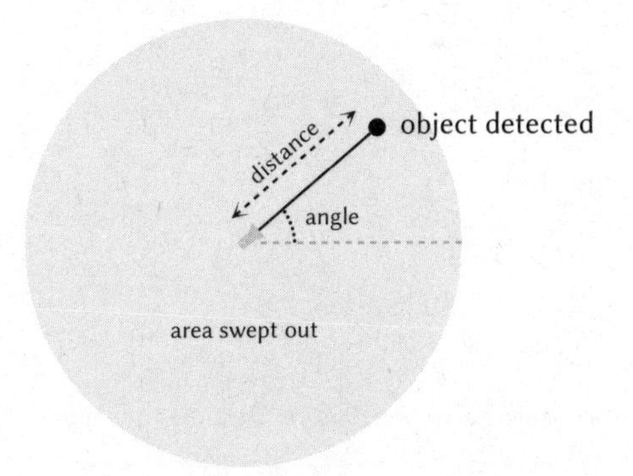

This is quite different from *x* and *y* coordinates in practice, but in the end we can achieve the same result, which is to precisely specify the position of any point in the plane.

Incidentally, cities that are on grid systems are essentially using cartesian coordinates. In Chicago it really is coordinates somewhat literally—the unit of distance is chosen so that there are 800 units to the mile, and there's a (0,0) position at the intersection of State Street and Madison Street downtown. Then "800 North Michigan Avenue" means a location on Michigan Avenue that is one mile north of Madison Street, and "400 West Randolph" means a location on Randolph Street that is half a mile west of State Street, and so on. This is very different from the system of numbering buildings in most places in the UK (and indeed other cities in the US, even with grids), where the buildings just go in order up the road, with odd numbers on one side and even numbers on the other. In Chicago you still have odd numbers on one side (the west of north–south streets and the north of east–west streets) and even numbers on the other, but not all house numbers exist, because the numbers are referring to coordinates rather than incrementally counting buildings.

However the system is much trickier in Sun City, Arizona, where some neighborhoods are in a circle:†

† Map images © OpenStreetMap contributors. Data is available under the Open Database License. See www.openstreetmap.org/copyright.

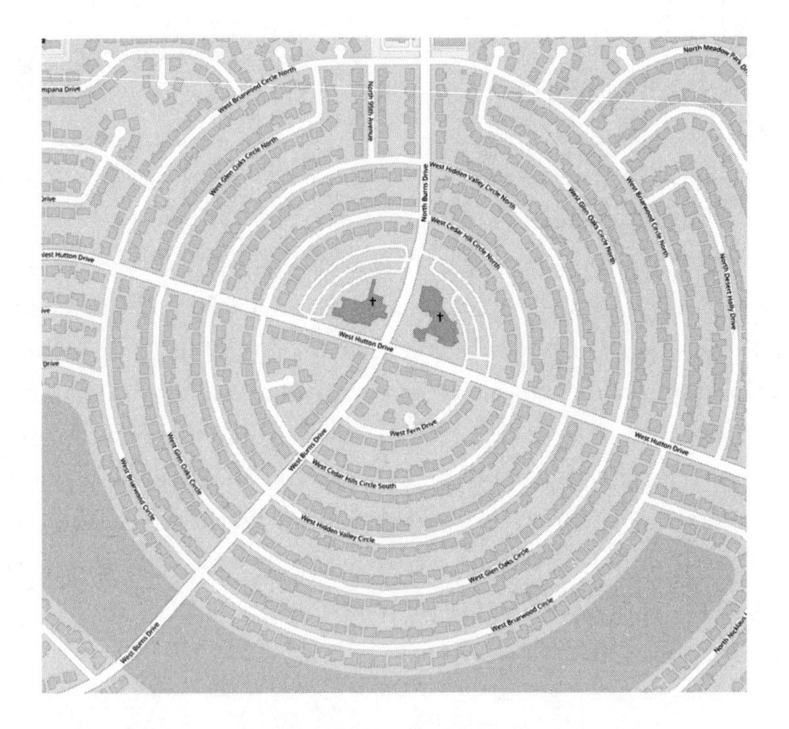

Amsterdam has a sort of polar coordinate "grid," with the canals (and hence the streets between them) running in concentric semicircles.

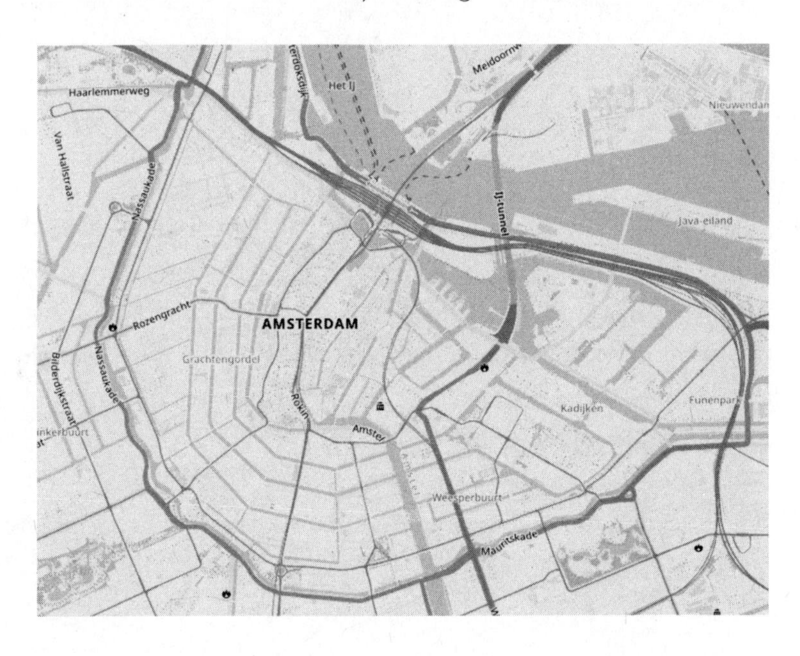

As with my "triangle" route through Cambridge that I mentioned in Chapter 1, these aren't geometrically perfect semicircles at all (unlike the circular streets in Sun City, which are much more modern and hence much closer to perfect circles), but it still makes sense to me to refer to them as "semicircular."

What I'm hinting at here is that there are many different ways to describe positions in two-dimensional space, so if we're going to try and describe straight lines in two-dimensional space, we first have to think about what system of coordinates we're going to use. If we're using polar coordinates then some straight lines are going to be much easier to describe than others. The radius lines (like the spokes of a wheel) arise rather naturally, as they are the path you take if you change your distance from the center but never change your angle. However, a line like this is not going to be very natural at all:

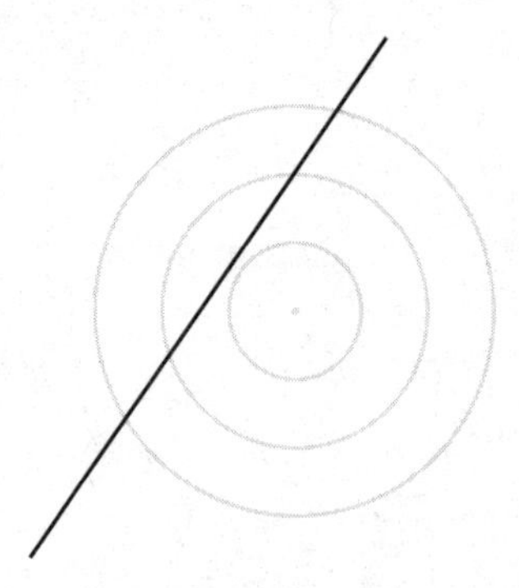

In this world, circles are much more natural than straight lines like that one.

So a first answer to "Why does $y = mx + b$?" is, as usual, that it's better to ask *where* it's true rather than *why* it's true: it's not always true, even once we've clarified that we're trying to describe a straight line. We need to be more specific and say we're doing it in cartesian coordinates.

The next thing to do is to understand what the equation is trying to tell us.

How a picture is described by an equation

The relationship between pictures and algebraic expressions is profound and remarkable, and we'll come back to it in Chapter 7. It's amazing, really, that we can use a string of symbols to draw a picture. (I used a formula to draw the text divider at the end of the previous chapter.)

Here's how it works. We're imagining drawing a straight line in a cartesian plane:

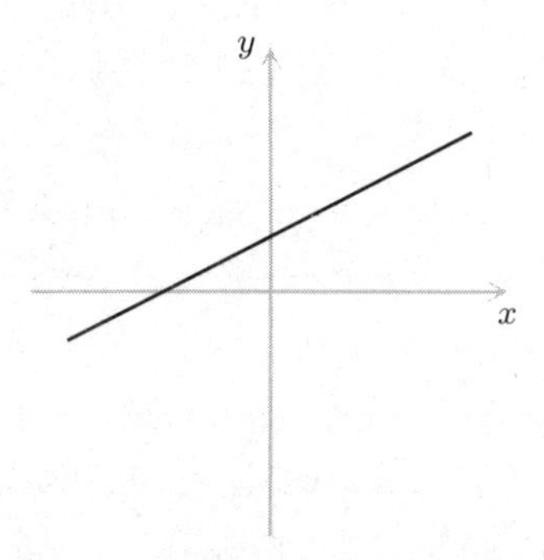

Now, every point on that line has an x coordinate and a y coordinate, and the question is whether there is some way to know in advance whether a given pair (x, y) is on that line. If we take *all* pairs (x, y) we'll get the entire plane, which is more than what we want. So what we want is just some particular pairs (x, y) and we want some way of saying precisely what they are without having to list them all, because there are infinitely many and so we physically can't list them all.

This is where those letters come in: not only can we not list all the pairs of coordinates for one particular straight line, we have even less hope of being able to list them all for *all possible* straight lines, because there are infinitely many lines as well as infinitely many points on each one. So instead we do that miraculous thing of "turning the numbers into letters" so that we can, in effect, express all of those infinite relationships all at once. This is the whole point of using letters instead of numbers.

Now what about these specific letters $y = mx + b$? The idea is that m and b are "constants," that is, once we've fixed a straight line, the numbers m and b stay the same. If we change them, we get a different straight line. So each m and b together pin down a straight line. Then, once we've fixed m and b, the equation gives us a relationship between x and y defining the x and y coordinates of all the points on that line. That is, any pair x and y satisfying that relationship gives a point that is definitely on the line, and also every point on that line is a pair x and y satisfying the relationship.

So using letters instead of numbers enables us to pin down infinitely many relationships for each of infinitely many lines. I don't think we spend enough time gazing in awe at how amazing that is. I urge you to take a moment to do it now. This is something we completely miss if we're really busy saying how obvious things are and discouraging "stupid" questions.

Now we've discussed the idea behind this formula, but not the specifics of the formula itself. One way we could investigate it is just to try a few examples. Remember we're fixing m and b and exploring what sort of x and y coordinates we get out of this formula.

We could try fixing $m = 1$ and $b = 0$ and then we'll get pairs of x and y coordinates such as:

x	$mx + b$
1	1
2	2
3	3
4	4

and if we plot those on a graph we'll get this:

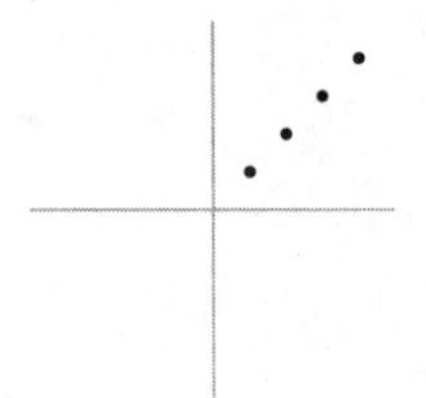

and we might go "Ah yes, that looks like a straight line," and fill it in.

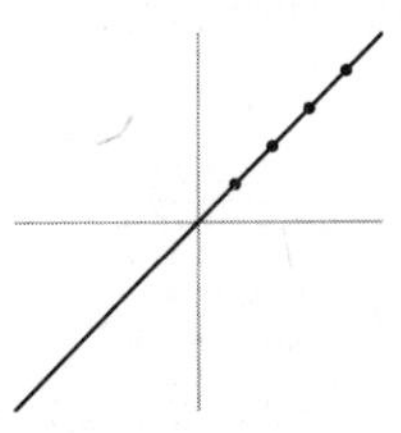

This would not be very mathematical because we have just taken a few examples and then extrapolated more or less by "psychic powers" rather than by logic; indeed there are plenty of possible graphs going through those four points, like this one for example:

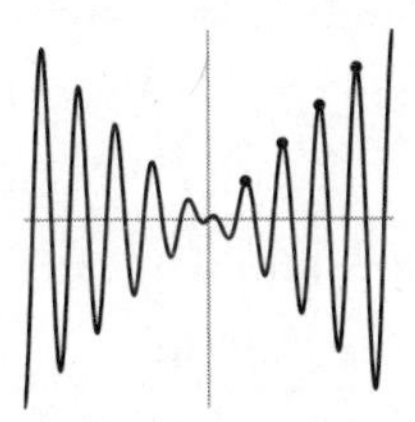

or indeed this much less organized but still perfectly valid one:

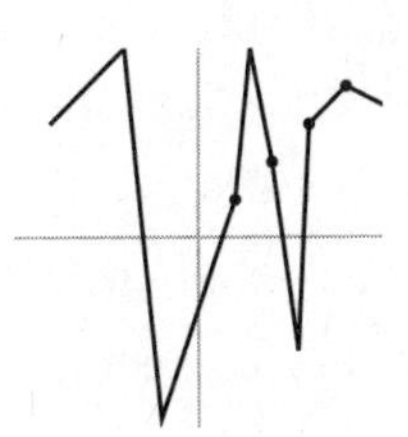

It would be more mathematical (that is, fully logical) to say: if $m = 1$ and $b = 0$ then the relationship "$y = mx + b$" becomes just $y = x$ (by substituting those chosen values of m and b into the equation). Now, which points on the cartesian plane satisfy that relationship? We need to ask ourselves which points have coordinates satisfying $y = x$, and the answer is: all the points where the horizontal distance is the same as the vertical distance, which is all the points on the diagonal line we first drew. Now instead of guessing, or using our psychic powers based on a small number of sample points, we have understood it logically.

We could now try all this for another example of m and b, say $m = 2$ and $b = 1$. We'd get this table and this attempt at guessing a line for the dots:

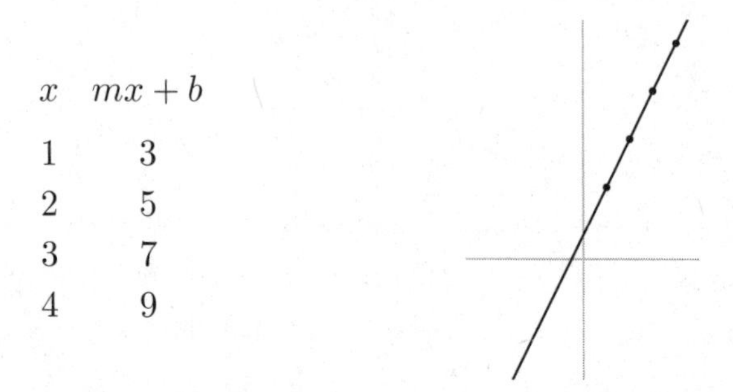

x	$mx + b$
1	3
2	5
3	7
4	9

and maybe we start noticing some trends, like the bigger m is, the steeper the line is, and the bigger b is, the further up the page the line is. We might even try some negative numbers, and realize that if m is negative it slopes downward, and if b is negative then the line is lower down the page rather than further up.

After trying a few of these we might feel it's "obvious" that this formula always gives a straight line, but that is not mathematical. Spotting those trends is not mathematical, it's more experimental, and trying a few examples does not logically explain to us the full situation. Spotting those trends and guessing the general pattern is the start of a mathematical discovery, but to make it math we need to make it a logical argument and not just a guess.

And this is where the question about straight lines is really profound, because how do you prove that *every* straight line has the formula "$y = mx + b$"? Well, as you might have started to guess based on some of our previous answers, aside from pinning down what context we're talking about, it also comes down to defining what a straight line is in the first place. That in turn comes down to making very careful and precise definitions of what geometry is. Mathematicians spent centuries trying to do that and eventually realized that there were way more types of geometry than they previously thought, and that straight lines only have the formula $y = mx + b$ in one type of geometry, and in other types of geometry they have different formulas. So it turns out that this equation is not true everywhere, not at all.

When straight lines don't look straight

How do we define a straight line? This is a deep question. One way to do it is to think about pulling a piece of string completely taut, or to think about how light travels: it always takes the shortest possible path. When we pull a piece of string taut, it takes the shortest path it can between the two endpoints in that particular space. So if we're trying to pull it taut around the corner of a building, it will take the shortest path in the space that takes into account the obstruction of the building. Here's a view from above, of a piece of string being pulled taut between two points, around the corner of a building. It's very hypothetical of course, and not necessarily realistic, but the point is to get an abstract idea of the principles.

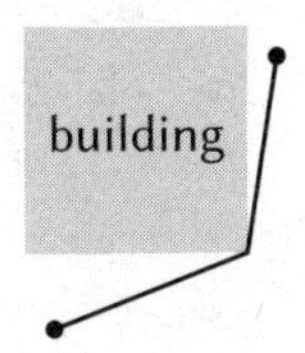

And this is the thing: the shortest path between two points depends on the shape of the space you're in. Moreover, it also depends on what notion of distance you're using. It is exceedingly contextual.

For example, "taxicab distance" is the type of distance that imagines we live in a city built on a grid, like downtown Chicago, and we can only go along roads. In that case the shortest path between the points shown on the map below is 7 blocks, because no matter which way you turn you will have to cover 3 blocks east and 4 blocks south to get there:†

† Map image © OpenStreetMap contributors. Data is available under the Open Database License. See www.openstreetmap.org/copyright.

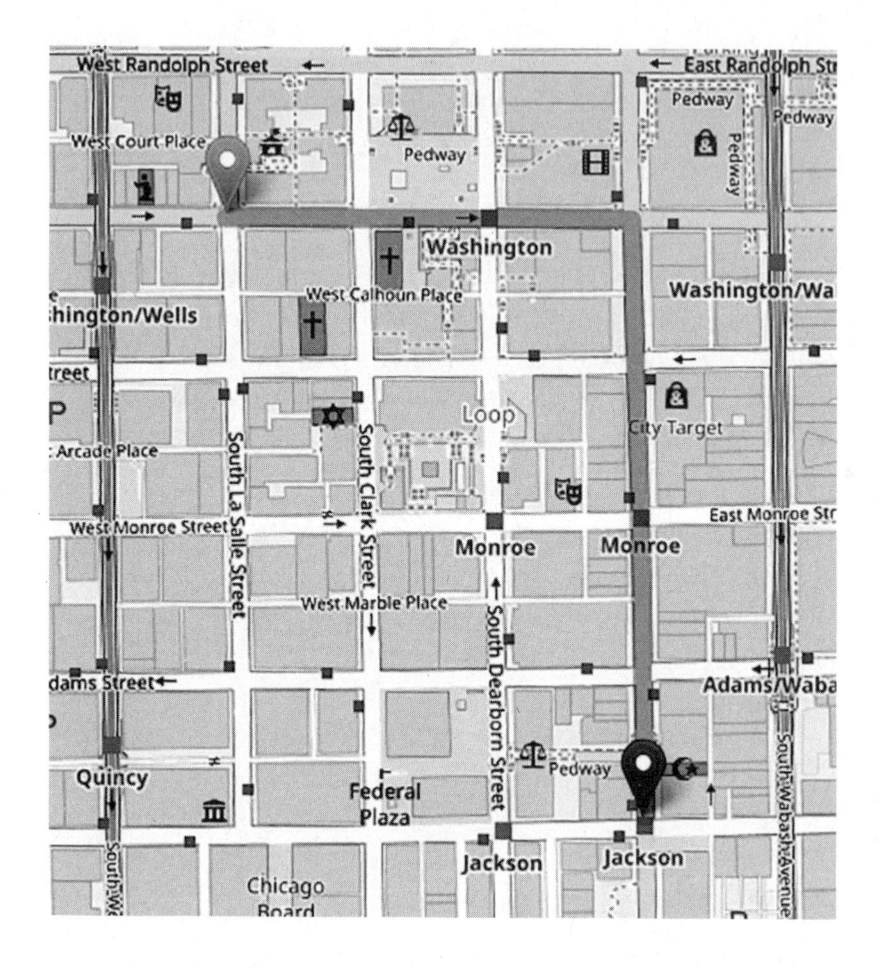

This means that any of the following would count as "straight lines" in this geometry, because each one is a shortest path from point A to point B, provided we don't count turning a corner as some extra arduousness.

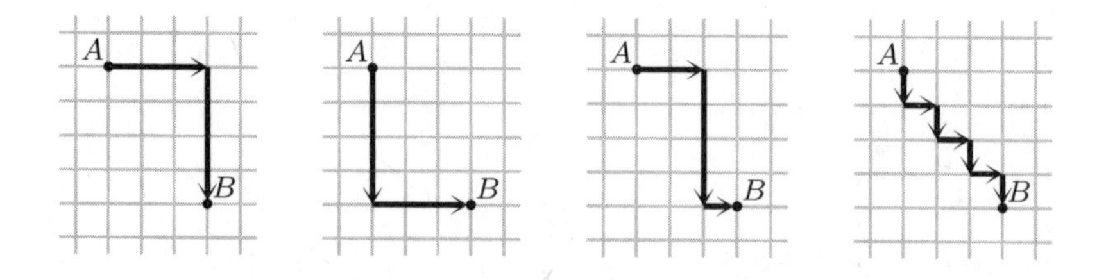

This is very different from the straight line that a crow could take flying from A to B, which would be a diagonal. This is a sense in which straight lines depend on the geometry of your world.

In 2021, it was spectacularly reported that some optical illusions were occurring on the English coast in which ships looked like they were floating above the surface of the sea. This is called a "superior mirage," "superior" not because it's better, but because the object looks higher than it is, as opposed to an "inferior mirage" where it looks lower than it is. It happens when there is warm air sitting on top of colder air, and as the colder air is denser, light travels more slowly through it. This causes the light's path to bend downward as it travels to our eye: but that is in a sense the "shortest path" for the light in the geometry created by the variable density of the air. The thing is that our brains are not that intelligent (I don't mean to insult them, but they do have limitations) and so they interpret the light as if it had traveled in a straight line in a rather basic notion of space, not taking into account the different densities. Thus it looks like the boat is hovering in the sky.

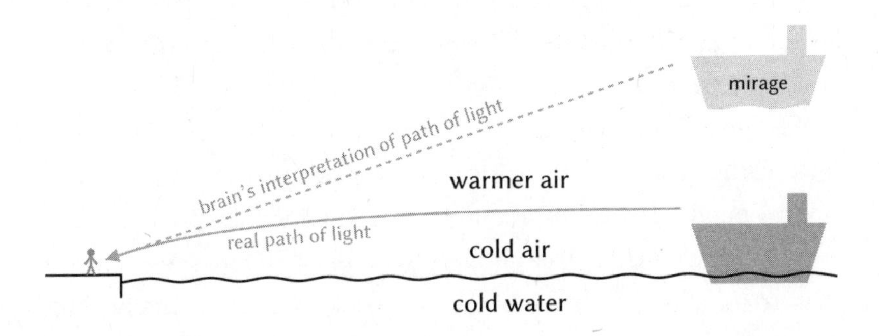

Imagine if someone out of sight lobbed a ball to you—it would curve through the air and approach you from above, like the beam of light in the diagram above. If your brain thought that balls could only travel in straight lines, you would think that the person must have been hovering in the sky and had thrown the ball to you from there. In reality, balls travel in curves because of gravity, and one of the key ideas of Einstein's theory of relativity is to think of those curves as straight lines in some other geometry, a geometry that takes the pull of gravity into account. Then we only see the curve of the ball as a non-straight line because we're not using the appropriate geometry.

The shortest distance between two points also doesn't look like a straight line on a globe shape such as the earth. It does look like a straight line if we're not going very far, because the earth is large enough that very small portions of it look pretty flat close up. So the shortest distance between two points that aren't too far apart still looks pretty much like our usual idea of a straight line. However, if you look at flight paths you might sometimes be surprised at how curved they look. They're not always exactly taking the shortest path (not through plain space anyway, because they also take into account currents and things), but they are basically trying to take the shortest path from A to B to save time, fuel and money. I periodically marvel at how far north the flight path from Chicago to London goes. I vaguely imagine going across the middle of the Atlantic, because Chicago is further south than London, but in reality the shortest path goes way up over Canada and practically Greenland. The shortest path from A to B in a particular type of geometry is called a geodesic.

The realization that there are different possible types of geometry took mathematicians by surprise. They were busily trying to understand Euclid's postulates about geometry, which are basic facts he thought were indelibly true about straight lines. But in trying to prove that straight lines could not be any other way, they accidentally found whole new types of geometry in which straight lines behave in a different way.

One of the new types of geometry they realized could exist is the geometry on the surface of a sphere, called, wait for it, spherical geometry. I think of it as "bulbous" geometry, because everything bulges out. If you think about trying to draw a triangle on the surface of a sphere, it is destined to bulge out somewhat. (I find it oddly satisfying to do this in ballpoint pen on an orange.) This "bulging out" can be encapsulated by looking at the sum of the angles in a triangle. In our normal "flat" (non-bulging) geometry, the angles in a triangle always add up to 180°. However, on the surface of a sphere, they will add up to *more* than 180°.

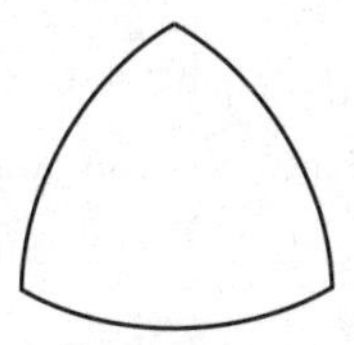

You might wonder if there's also a "bulging in" geometry rather than one that "bulges out." Perhaps you can even vaguely dream it up in your head. If so, you're thinking like a mathematician. And at this point you might guess that triangles in this geometry have angles that add up to *less* than 180°. That sort of geometry is called *hyperbolic geometry*. It's a little harder to imagine, but maybe not if you've ever done any sewing, or knitting, or crocheting. If you want to crochet a flat circle, say for a coaster or place mat, then you might start in the middle and work your way outward in concentric circles. You have to carefully increase the number of stitches on each circle by just the right amount so that the whole thing stays flat. Now, if you do fewer stitches than that, you will end up making a bowl shape, because the whole thing will draw in instead of staying flat. If you do *more* stitches than the flat version, then you will get too much material at the edges and it will sort of look frilly. That's a bit like what hyperbolic geometry looks like. A simpler version of that is a saddle (for a horse) or indeed a Pringle.† If you imagine drawing a triangle on a Pringle, by starting with three dots and then joining them up with the shortest possible lines between them, it will look a bit like this, a skinnier version of a triangle, or a triangle "bulging in":

† The "stackable potato-based chip," which is made stackable by its geometry.

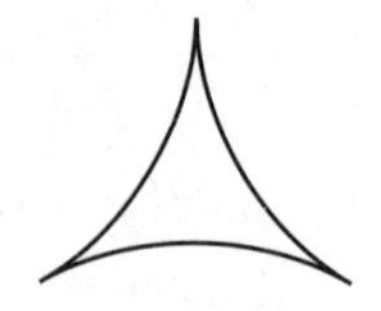

So really, it's not absolutely true that a straight line has the formula $y = mx + b$, it's only true in one particular type of "flat" geometry. In fact it's practically a definition of that particular type of geometry, the type that Euclid was originally trying to characterize by thinking really hard about straight lines. That type is now called Euclidean geometry.

What's the point?

The point of doing all this with letters is to express more things at once. It's about building up techniques so that we can get further with our reasoning, and reason about more complex things. It's also about having greater transferability, so that we can use what we've understood in a wider range of places.

In the example of diagrams for factors of numbers, we went from a diagram of factors of 30, to the understanding that this comes from a product of three different prime numbers a, b, c, to the understanding that this is really about any set of three things a, b, c. Once we are at that level of abstraction, the ideas are far more broadly transferable than when we were just thinking about the factors of 30. It's another form of indirect usefulness. Is it "useful" to understand the factors of 30? I would say, not terribly, not directly anyway. But the full understanding of that situation and the light it then sheds on social structures is very useful, in an indirect way.

One big complaint about math is that it's pointless and you never end up using it in the rest of your life. Unfortunately if you've only been told about direct usefulness and you've been led to believe that math is all about direct usefulness, then you might well be right. I don't think it is directly very useful to know that the equation of a straight line in two-dimensional Euclidean geometry, expressed in cartesian coordinates, is $y = mx + b$. I'm certain I've never used it in my daily life. What definitely is useful is the amount of practice my brain has had at carefully exploring different worlds and adjusting individual constraints to observe the consequences. For me that is the point of abstraction, it's the point of using letters to refer to numbers, and it's the point of the exploration of straight lines in different types of geometry.

If you understand how to manipulate unknown quantities *in general*, then that is much more likely to be transferable than learning to manipulate specific individual things. So there are two different questions here. There's the question of why something is a fruitful technique for a mathematician, who may well use that specific technique in future work. And then there's the question of why it is in any way relevant to anyone else, who is probably never going to use that specific technique in the rest of their life.

My answer to that last question is the same as my answer to another question I'm often asked, which is how I come up with explanations and diagrams that bring clarity to various sensitive, delicate, nuanced and convoluted social arguments. The answer is that my training in the discipline of abstract mathematics makes those things come to me very smoothly. I couldn't say exactly *how* manipulating symbols leads to my ability to see helpful ways to use abstraction to understand the world, but it comes back to that idea of core training for the brain.

On another occasion I was taking part in a day of public talks on "Mathematics in life," with a series of speakers all of whom were applied mathematicians except me. There were wonderful talks on the math of gerrymandering, the math of error-correction on CDs,† the math of cryptography, and the math of chocolate fountains. There was also me, talking about logic and abstraction and political arguments. At the end of the day was a panel question time and audience

† CDs are somewhat obsolete now, but digital error-correction goes beyond CDs.

members were invited to submit questions to any speaker. One person asked all of us how we use our research in our actual daily life. All the applied mathematicians agreed that they didn't directly use their applied math research in their daily life, that it was really the techniques and discipline of mathematics in general that they use. For once I felt I could stand up for abstract math and say that those techniques and that discipline are the content of abstract math, and in that sense I actually do use my research in my daily life, just not in a way that you'd recognize if you're only focused on direct applications.

CHAPTER 6

FORMULAS

Where do all those trigonometric formulas come from? And why do we have to memorize them?

I hope you can start to feel what the answer to the second question is going to be: really we don't have to memorize them, if we understand where they come from. In the previous chapter we looked at the formula for a straight line, which enabled us to convey a picture rigorously as a string of symbols. We expressed the graph by finding something all the points on it had in common, in the form of a relationship between their x and y coordinate. In this chapter we're going to look at formulas giving relationships between things that each have their own entire graph: the trigonometric functions for sine, cosine and tangent. It can seem that formulas are just there to test us, but really they're there to help us. I want to continue showing what a formula is really, to me: an almost magic machine that enables us to do infinitely many things at the same time. The most wonderful formulas are the ones that actually explain something. Often it's in trying to explain things to someone else that we understand them better, and sometimes a formula is the most succinct way to do it; it's just that the succinctness can seem rather sudden and baffling if it comes out of nowhere. Powerful machines can always seem baffling if you're not used to them; imagine someone from two hundred years ago seeing a jumbo jet. Formulas are like powerful machines, and equations are like miraculous bridges enabling us to travel between different mathematical worlds.

Memorizing vs. internalizing

You might think that formulas are just definitions and so there's nothing to understand about them; this might lead to the notion that we just have to memorize them. However, definitions in math are motivated by something, and if we understand those motivations we have a chance of internalizing them rather than memorizing them, which is slightly but crucially different. Internalizing something is where it gets embedded into your consciousness by some combination of understanding, intuition, repeated use, familiarity. Memorizing, to me, is where you commit something to memory by sheer brute force, like rote memorization, or the use of a completely unrelated mnemonic like SOHCAHTOA for the trigonometric (trig) expressions "Sine equals Opposite over Hypotenuse, Cosine equals Adjacent over Hypotenuse, Tan equals Opposite over Adjacent," or FOIL for how to multiply parentheses out by "First, Outer, Inner, Last," or PEMDAS for the order of operations "Parentheses, Exponent, Multiply, Divide, Add, Subtract," or MR VANS TRAMPED for the French verbs that form the past tense irregularly:

Mourir	**V**enir	**T**omber
Rester	**A**ller	**R**etourner
	Naître	**A**rriver
	Sortir	**M**onter
		Partir
		Entrer
		Descendre

The latter is a good example of the difference between what I'm calling memorization and what I'm calling internalization, even if you don't speak French. (With apologies to a French translator if this is ever translated into French, because this part is going to be hard to translate into French, and now I'm just making it worse.) The mnemonic did enable me to produce a list of these verbs, having not been formally tested on them for some years. However, it has never helped me actually use the verbs in normal speech. You would sound very stilted if you stopped every time you got to a verb and ran through this list in your head, using the mnemonic, to see how to form the past tense. Rather, for actual use and fluency, you have to *internalize* the use of these verbs, so that through some combination of repeated use, familiarity and understanding, you know deep inside which construction is right.

This difference between memorization and internalization is important, but we don't make the difference clear enough, and so we get into contentious arguments about whether "memorization" is important in math, when everyone in the argument might mean something slightly different by that. I personally think there are very few times when rote memorization is important. It might help, but only if someone *enjoys* it. If someone doesn't enjoy it then the harm outweighs the benefits. The question of harm is dramatically under-considered when we talk about math education. Certain things may well help students do better in a math test, but we should consider if those things also cause so much math trauma that the student then shuns math for the rest of their life, in which case the net long-term result is that we didn't teach them anything much besides an aversion to math.

The funny thing is, I personally did rote memorize the definitions of trig functions when I was at school (not by SOHCAHTOA, which I hadn't heard of until I was a professor and a new generation of students told me about it). Only much later, when I was teaching, did I understand that there's a beautiful relationship going on here, and then I wished someone had told me about it earlier. I was relatively unscathed, because I didn't actually mind memorizing those formulas, but this is not the case for everyone.

The formulas I'm referring to are the ones relating the trigonometric functions sine, cosine and tangent to the sides of a triangle. You might remember what a sine wave looks like, and if we draw it on axes it looks like the graph below on the left, whereas cosine is shifted over a bit, as shown in the right-hand graph:

These days we can calculate sine or cosine of anything just using a calculator or computer, but the formulas tell us how to apply the functions to an angle in a right-angled triangle, using the sides of the triangle. The sides are described relative to the angle we're using, as in this diagram, where the angle in question is marked with the curved dotted line:

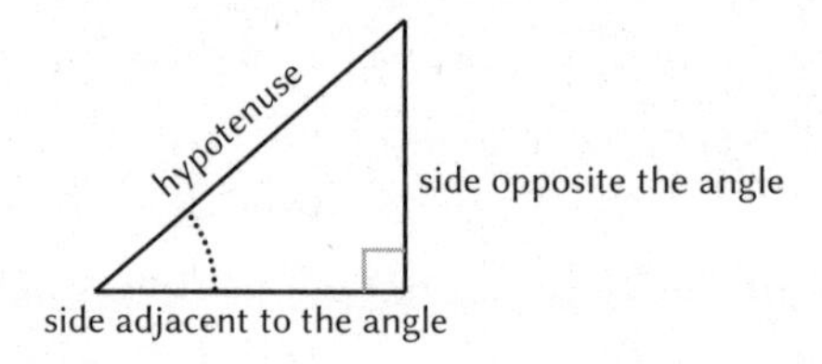

Now here's what the mnemonic is supposed to tell us:

- SOH: **s**in equals **o**pposite over **h**ypotenuse
- CAH: **c**os equals **a**djacent over **h**ypotenuse
- TOA: **t**an equals **o**pposite over **a**djacent

The reason I, along with so many other students, felt I had to memorize this is that nobody explained where those definitions came from, and so I had no other way to do it. Now that I understand where they came from, I don't need the rote memory version. Moreover it's been so long since I did the rote memorizing, I'm not sure I can produce the rote memorization correctly—which is another problem with rote memorization. By contrast, if you understand the real point of the formulas, you can reproduce them from your understanding, which is much more reliable than rote memory.

There could be different opinions about the "real point" of trig functions, but in my opinion the point is that they're explaining a relationship between circles and squares, or between circular grids and right-angled grids.

Circular vs. square grids

In the last chapter we talked about two possible ways of describing points in a two-dimensional plane: using cartesian coordinates (a right-angled grid), or polar coordinates (a circular grid). We also said it doesn't really matter what system you use for describing points in a plane, as long as you know how to translate your information between different places.

So here's the key: How do we translate between polar and cartesian coordinates?

Suppose we know the polar coordinates of a point, and we want to turn those into cartesian coordinates. What we know is the angle and the direct distance from the origin, as shown here:

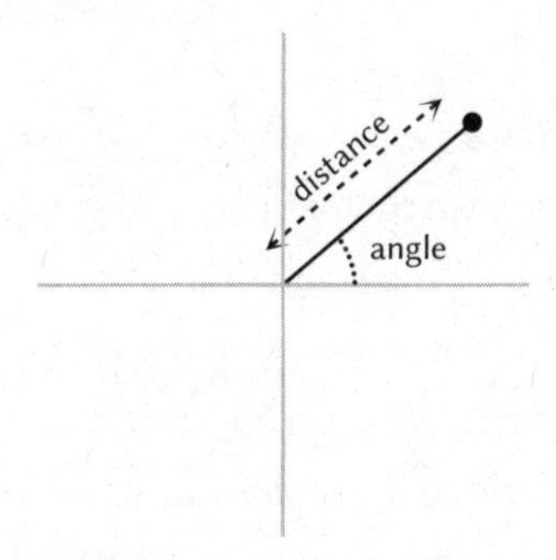

How do we turn those into an x and a y coordinate? Well that is talking about this right-angled triangle shown below:

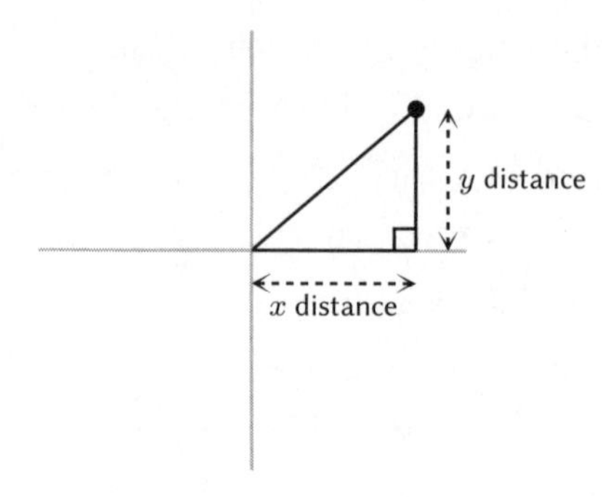

Note that it's right-angled precisely because we're trying to get ourselves into the right-angled cartesian framework. So we're starting with the information of the angle and the long edge of the triangle (otherwise known as the hypotenuse) and we're trying to produce the other two sides of the triangle.

To start with cartesian coordinates and turn them into polar coordinates we'd be starting with a point expressed as an x distance and a y distance, and trying to express it as an angle and a direct distance from the origin. So we're starting with the two shorter sides of the triangle, and trying to produce the hypotenuse and the angle (that is, one of the angles that isn't the right angle).

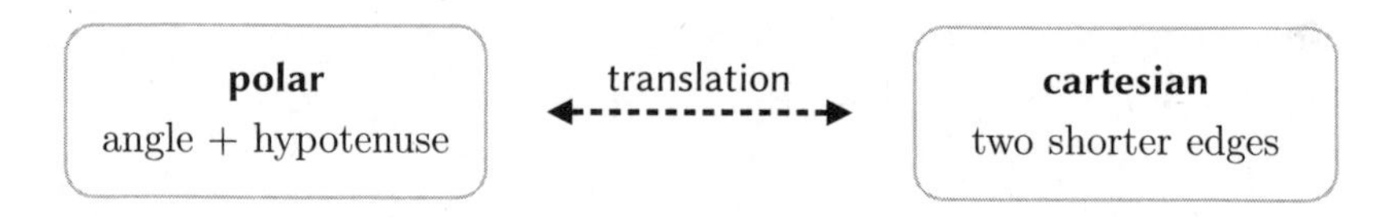

What I just described dealt with any particular point in the two-dimensional plane, but mathematicians like dealing with the entire situation at once, not just one point at a time. We want to understand the entire overall relationship between the two worlds, not just have a way to translate a point at a time. (Perhaps it's a bit like the fact that a language translator is much more than a dictionary.) If we think about the translation for varying points rather than for one point at a time, we can imagine going around and around a circle in the polar world, and watching how our x and y coordinates vary as we go around.

It's as if you're riding on a Ferris wheel, for example. As you go around, you might notice your vertical movement more vividly than your horizontal movement. After all, we humans move around horizontally quite a lot, but moving vertically is more novel. Your vertical movement will be very noticeable when you're at the side, going up. Then it slows down as you go over the top, and for a brief instant it will feel like you're stationary when you reach the top. Then your vertical motion will speed up again as you're going down the other side, and then it will slow down again as you go around the bottom.

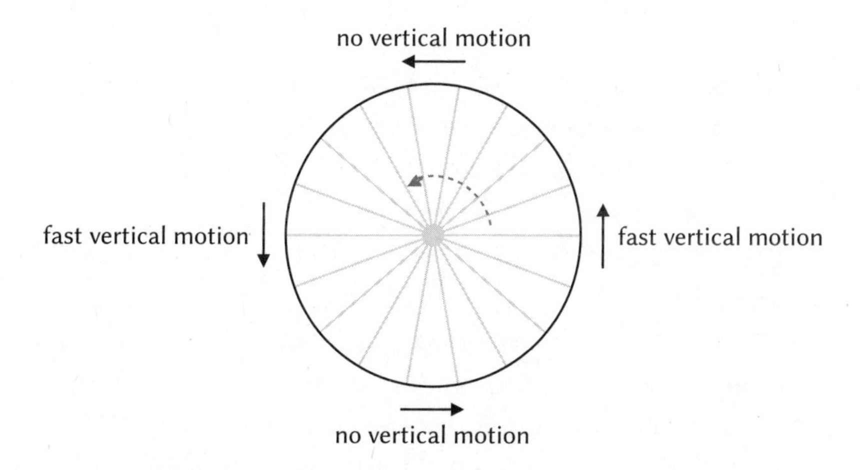

If you do notice your horizontal movement, it will have its drama the other way around: it is most noticeable as you're going over the top or "under the bottom," and slows down as you get to the sides, with a moment of horizontal stillness when you reach the extreme outer points and are traveling entirely vertically.

Now for the punchline: this is where the sine and cosine functions come from. If you only notice your vertical motion, that's the sine function, and if you only notice your horizontal motion, it's a complementary motion and is called cosine. Here's a picture of the sine function and how it corresponds to the changing angle as you go around a circle. The key is to remember (from our earlier diagram of how polar coordinates work) that the angle is measured relative to the horizontal

x-axis, so 0° and 180° are the sides of the circle, and 90° and 270° are the top and bottom.

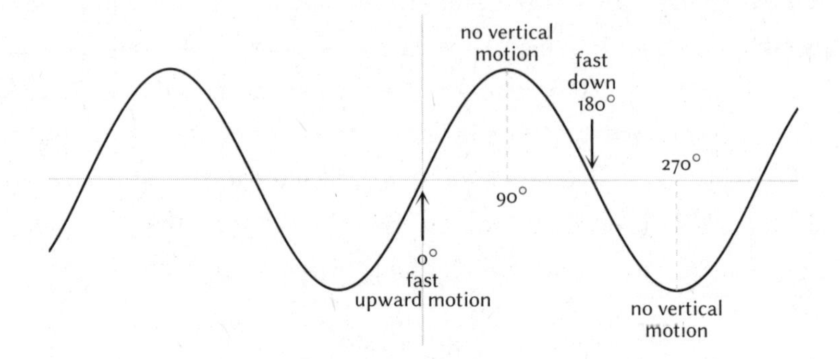

In math, the prefix "co" is often used to indicate something that is somehow complementary, which often means looking at the same concept from an opposite or reversed point of view. This is the sense in which sine and cosine are complementary.

You might still wonder why sine and cosine are that way around, and where the word "sine" comes from in the first place.

The use of the word "sine" seems to be a tale of linguistic misunderstanding. It appears to come from a Latin translation of a mistaken interpretation of an Arabic transliteration of a Sanskrit word. There are unfortunately many examples of words creeping into English that have been appropriated from other languages not quite correctly. For example "chai" really just means tea, so calling something "chai tea" is like saying "tea tea." On the subject of tea, Westerners who like Chinese food seem to speak of going for "dim sum," which is all very well but in Hong Kong they say they're going for "yum cha," which literally translates as drinking tea, but really means eating dim sum (usually accompanied by tea). There are also many immigrants whose family history is hard to trace because of different transcriptions of their name that were used, not to mention abbreviations, simplifications, misunderstandings or all-out acts of erasure.

The study of trigonometry in general goes back far into ancient cultures, but the sine function as we now know it was really developed by

Indian astronomers in the fourth and fifth centuries. The first known reference is by Aryabhata the Elder, who used the word *jya* in Sanskrit meaning "bow-string" (as in a bow and arrow) to refer to a half-chord. In this context, a "chord" is a straight line joining two points on a circle, which is somewhat like a bow-string. From the picture below, we can see that the chord length is double the vertical distance if we take the center of the circle to be at 0, and rotate our circle so that the chord is vertical.

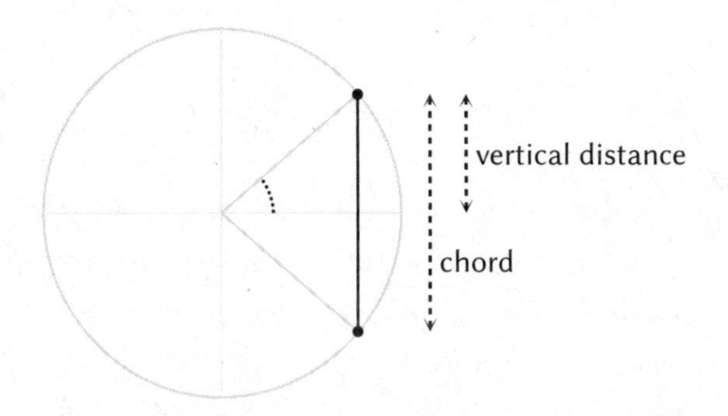

Of course, if you're thinking of a Ferris wheel then the center isn't at ground level, but perhaps we could imagine a Ferris wheel that goes half underwater. We'd need waterproof enclosed capsules, mechanisms that work underwater, all sorts of safety measures... Well this is one difference between math and engineering: I can sit here on my sofa dreaming of a half-underwater Ferris wheel and not worry about those engineering details or whether the thing is at all realistic. Now that I've thought of it, I rather like the idea though.

The reason we take the center of the circle to be at 0 in both directions is for convenience, or tidiness, or because we're lazy, or because we want to remove extraneous complications from the situation (maybe all those are the same thing). Sometimes math is like doing a controlled experiment, where you want to home in on one particular aspect of a situation, so you set it up so that the other aspects don't interfere. In this case, once we've understood how this works with the center of the circle being at 0, it's not too hard to work out what happens if the center is shifted.

There's also some pleasing symmetry to the center being at 0, plus it takes us back to the idea of polar coordinates, where the whole point is to have concentric circles around 0. We could try placing the circle in either of these ways:

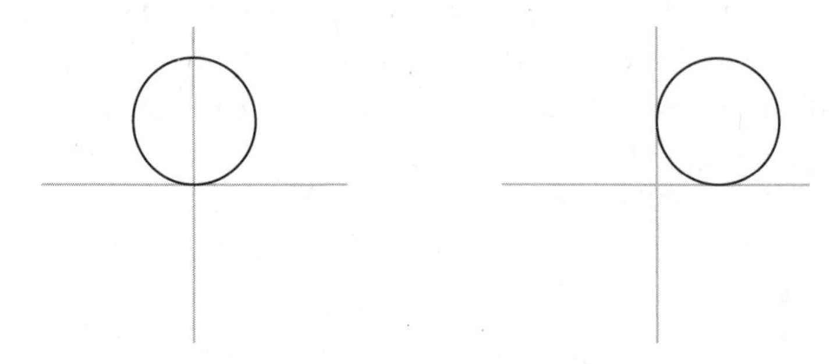

but it would add complications without adding any insight. If something adds complications *and* insight, then it's worth sitting down and weighing those things up, but in this case there's not much weighing to do.

Another thing mathematicians often do to make a controlled experiment is to scale the situation up or down so that as many numbers as possible become 1. For the circle, it's awfully convenient if we declare that the radius of the circle is 1. We can then understand all other circles in relation to that one, by scaling up or down.

You might wonder "1 what?" as in, what units of measurement are we using? As I mentioned earlier, this is another thing I really like about pure math as opposed to engineering, or even physics: it doesn't matter what units of measurement we use, so we don't have to mention them. We just assume we're always using the same units inside a given scenario. It's like how we do numbers in the first place: we don't always have to be specifying that we're adding 2 of this with 3 of the same thing, we just say 2 plus 3 and understand it to mean that we're taking 2 things and 3 things. It's also like when a recipe is specified by ratios, such as when you make oatmeal by taking one part oats to two parts water (by volume). It doesn't matter how big one "part" is, as long as you take the same size "part" for your water.

Like many people, I had some measurements "trauma" at school. I recently found an old physics test in which we were supposed to read a graph, and then the question said "How many seconds does it take the dog to reach the ball?" or something like that. I wrote "5" and lost half a point because I didn't say "seconds," despite the fact that the question said "how many seconds." As you can probably tell, I'm still a bit miffed about that. It's the kind of thing that can put students off a subject forever.

Anyway, with the proviso that the center of our big wheel is at height 0, and that its radius is 1, then the vertical height is the sine function.

Sine and cosine

Now that we're ready to talk about the trigonometric functions, I feel the need to start using some letters to refer to all these varying numbers. Mathematicians often like using Greek letters for angles and Roman letters for lengths of edges, to remind us that the letters are playing slightly different roles. I will use the Greek letter θ (*theta*) for the angle, while using the usual x for horizontal distance and y for vertical distance. What we're saying is that there is a fixed relationship between θ and y called sine, and it can be expressed like this:

$$y = \sin\theta$$

Yes, the word is sine but mathematicians abbreviate it down to the three letters sin.

Right at the start of all this we were investigating how to "translate" between polar coordinates and cartesian coordinates, and so far we've investigated how to express our y coordinate. The x coordinate also has a fixed relationship with the angle θ, and that's what we call cosine. We declare

$$x = \cos\theta$$

One thing you might notice about this is that sine and cosine aren't *really* different. That is, because of the symmetry of a circle, horizontal motion and vertical motion really operate with the same pattern. Our notions of horizontal and vertical are quite arbitrary (well, maybe they're to do with how we feel gravity), and we could pick any other reference lines and get the same sorts of patterns. Symmetry often gives us clues to things. In this case it can help us understand that cosine is the same as sine really, just shifted a bit. Sine goes really fast at the sides and slowly at the top and bottom, whereas cosine goes fast at the top and bottom but slowly at the sides. If we rotate our frame of reference (say, we lie on our sides and look at the situation sideways) those two things will switch role. That's why the graphs of sine and cosine look so similar.

One possible explanation (emotionally anyway) for why we have chosen them this way around is that there's something satisfying about measuring both the angle and the distance from the same reference axis.

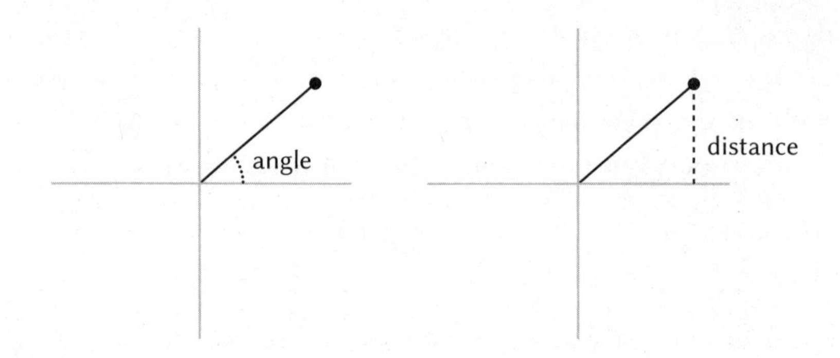

This means we can start with zero angle and zero distance, and then both of them start increasing. Whereas if we measured angle from one axis, but measured distance from the other axis, then we'd start with zero angle but maximum distance. Really this isn't about favoring vertical over horizontal, it's about starting all our measurements from the same axis. So we take this as the more "fundamental" relationship (sine) and the other is the complementary one (cosine).

Incidentally, this has not yet told us anything about how we actually know the value of the sine function for any particular angle. That is a difficult problem involving calculus, but we can get an idea of it by looking around us a bit. Actually this happened to me at a conference once when we were given wraps for lunch. They were cut on the bias, which is often done to expose more of the filling and make it look like there is more in there—if you cut at an angle, you get a bigger cross-section than if you cut straight down.

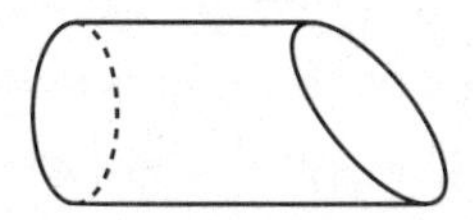

On this occasion I thought there was too much wrap for me and wanted to remove a layer, so I unrolled the wrap and removed one layer. And lo, there I was, holding a sine wave.

I'm afraid I then missed most of the rest of that talk because I had to do a little back-of-the-(literal)-napkin calculation to check that I wasn't imagining it, and that really is a sine wave.

In the olden days of my bygone youth I used to spend a lot of time fiddling with spiral telephone cords while on the phone: looping them around, stretching them out, trying to get rid of kinks. If you stretch out a spiral phone cord, or indeed any other sort of spiral (such as a Slinky) then what you can see from the side is a sine wave. That's because if you stretch it out and look at it from the side, you only see the vertical coordinate as the cord goes around and around; you just see it going up and down as our eyes don't really see it going away from us and toward us. To see just the horizontal coordinate, you'd look from above instead of from the side. But the symmetry means it won't look any different, just shifted along a bit. Here's a picture of a Slinky I have stretched out. The trouble with a photo is that the lens is fixed in the middle, so it sees the center of the Slinky straight on, but it sees the sides from a slight angle, enough to make it not quite a sine wave anymore. But I hope you can see that at least one part in the middle looks a lot like a sine wave.

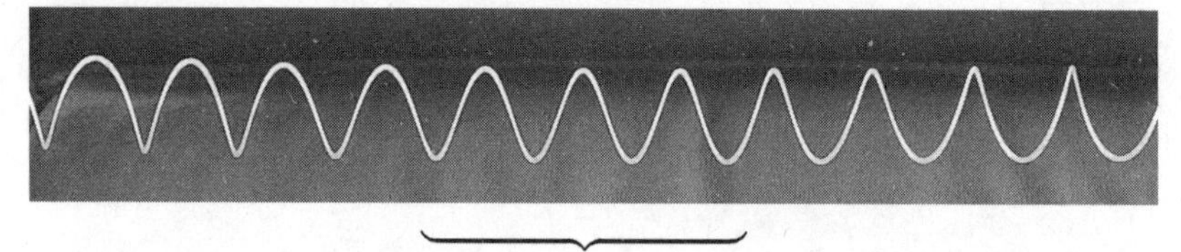

looks quite like a sine wave

I once tried making a sine wave by taking a long-exposure photo of me walking along and waving a LED light around and around in a circle. I walked perpendicular to the direction of the camera, and waved my hand in a circle in front of me, in the direction up–right–down–left–up–right–down–left:

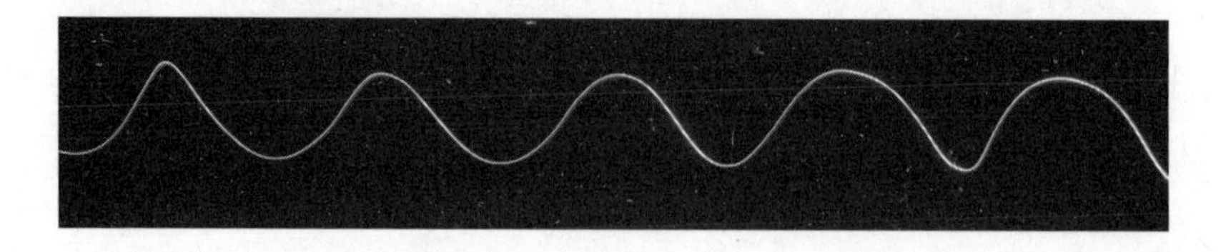

It was hard to make sure I was walking at a constant pace and also waving my arm at a constant pace, but the result was definitely at least vaguely reminiscent of a sine wave.

The fact that we're going around and around a circle shows us why the sine function repeats itself periodically, which is indeed called "periodic" in math: when we come back to the beginning of the circle the y coordinates will just start repeating themselves again.

Drawing pictures of the circle and the x and y coordinate, and thinking about a little geometry, can then help us understand some of the relationships between trigonometric functions. We can then express those relationships as formulas.

Relationships and formulas

Here again is a picture of a point on a circle, and its x and y coordinates, together with the right-angled triangle that we're implicitly thinking about in that case. Again we will write θ for the angle as shown in the diagram, so then the y distance is $\sin\theta$ and the x distance is $\cos\theta$.

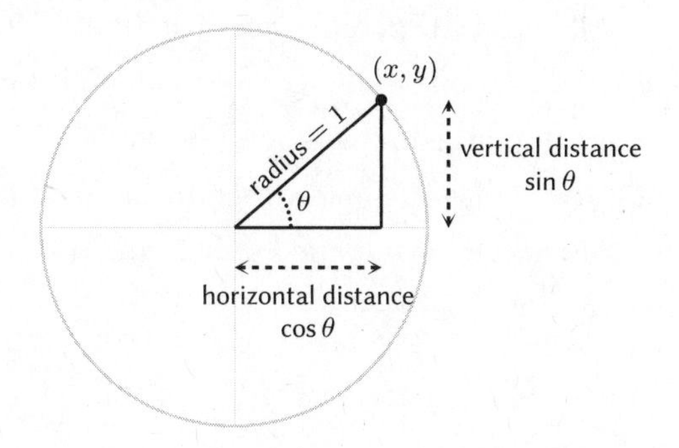

One immediate result is that we can use Pythagoras's theorem, which says that $a^2 + b^2 = c^2$, where a and b are the shorter sides of the triangle (necessarily the two that meet at a right-angle) and c is the long side, the "hypotenuse."

In this case a and b are $\sin\theta$ and $\cos\theta$, and the hypotenuse is 1. So Pythagoras gives us this relationship between sine and cosine:

$$\begin{aligned}(\sin\theta)^2 + (\cos\theta)^2 &= 1^2 \\ &= 1\end{aligned}$$

If we think about the geometry of scaling triangles, we can understand a little bit more of the infamous mnemonic SOHCAHTOA. We do still have to understand one basic principle, which is the principle of scaling things up and down: if we scale something up or down but keep its shape, what we're doing is leaving all the angles the same, but multiplying every length by the same number, a "scaling factor." So for example if I start with the triangle below on the left, I can leave every angle the same but multiply every length by 2 and get the one on the right.

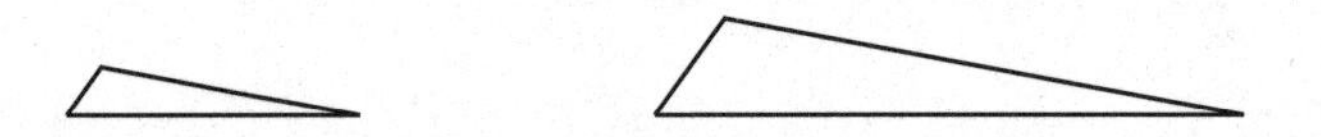

If I start with any shape, no matter how complicated, I can do the same thing: as long as I multiply every length by the same scaling factor I can leave all the angles the same, which means the picture will look the same to us, just at a different scale. I actually draw these pictures using that very principle in the code, because all I have to do is change the basic unit of measurement, and leave the rest of the code the same. So for the first diagram below I set the basic unit of measurement to be 1 mm and in the second one I set it to be 2 mm.

Deep down, it's as if we are viewing the same picture from a different distance—it just *looks* different to us, but it's the same picture in both cases. Scaling is definitely about *multiplying*, not adding—if I tried adding 10 to each length instead of multiplying, the triangle would transform as shown below: the angles change, so the shape changes.

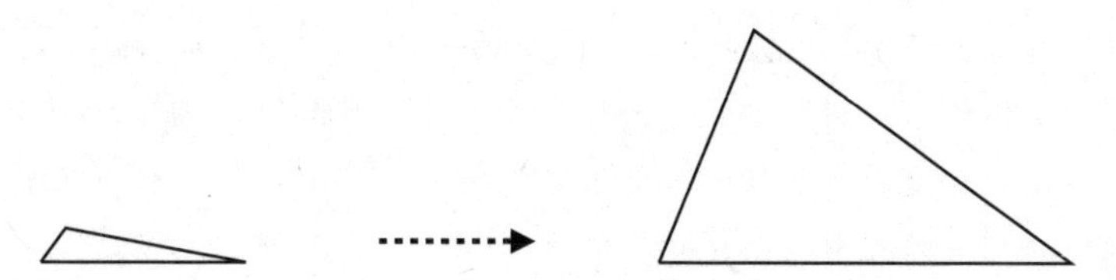

The principle of scaling means I can start with the diagram I've understood for the unit circle (radius 1) and scale it up to any size I want. For example if I want to understand radius 2 instead, I get this triangle:

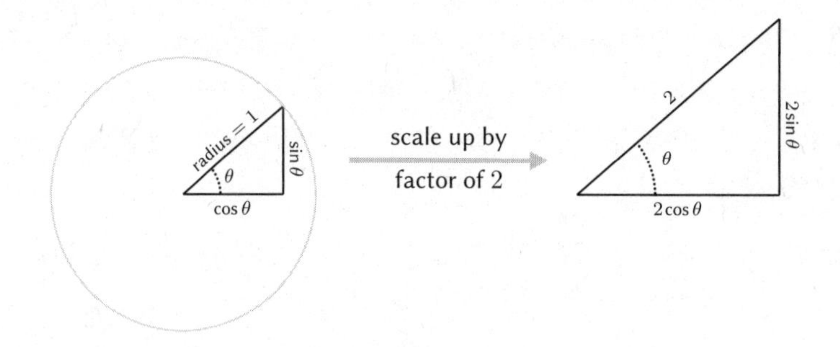

Now we see that as I've multiplied the radius by 2, if I want to keep the triangle having the same overall shape (same angles) I need to multiply every length by 2 as well. This means that the y coordinate is now $2\sin\theta$ and the x coordinate is $2\cos\theta$.

We can now invoke our letters and say that we don't want to list this relationship for every possible size of triangle, so we can say: suppose we call the length of the radius (hypotenuse) h. Then the whole triangle needs to be scaled by h. So the y coordinate is now $h\sin\theta$ and the x coordinate is $h\cos\theta$.

The final step in all this is to understand that sin and cos apply even if the triangle is pointing another way so that the "y coordinate" is no longer vertical and the "x coordinate" is no longer horizontal. For example if we turn the triangle any of these ways up.

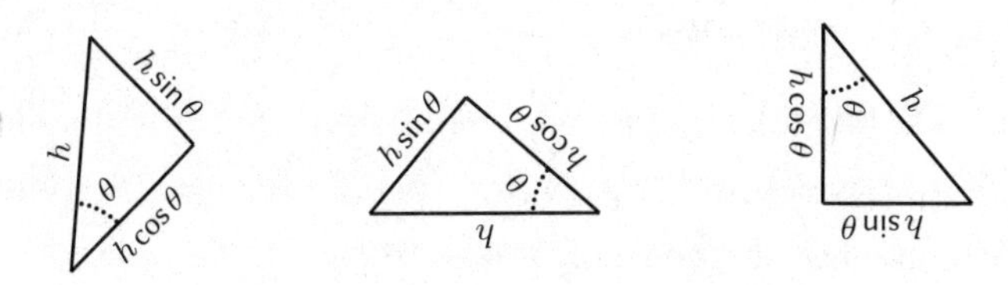

The general principles hold as long as one of the angles is a right angle, which means that it is possible to rotate the triangle so that we have a horizontal edge and a vertical edge, with the long edge (the hypotenuse) looking like a diagonal.

In order to avoid actually having to rotate our triangles before thinking about them, it's better for us to have a more robust way to refer to the y coordinate and x coordinate, that doesn't depend on which way up our triangle is. This is why we think of the y coordinate as "the side of the triangle *opposite* the angle we're thinking of," and the x coordinate as "the side of the triangle *adjacent* to the angle we're thinking of." It's a question of ruling out some potential ambiguity.

We've now said that

$$\text{opposite} = h \sin\theta$$

which means, if we rearrange things a little,

$$\sin\theta = \frac{\text{opposite}}{\text{hypotenuse}}$$

and

$$\text{adjacent} = h \cos\theta$$

which can be rearranged to produce

$$\cos\theta = \frac{\text{adjacent}}{\text{hypotenuse}}$$

And lo, these formulas can be rote memorized as "SOH" (for sine, opposite, hypotenuse) and "CAH" (for cosine, adjacent, hypotenuse). We've arrived at the first part of "SOHCAHTOA."

That leaves the last part, "TOA," which is talking about the tangent or tan function. This one is a little different (and not so profound) because it's just talking about the slope of the "spoke" or radius we're on. We can measure how steep a slope is by thinking about the rate at which it goes up as we move along horizontally. This is how slopes are expressed in road signs. If you see a sign saying 1:5 then it means the vertical distance changes by 1 unit for every 5 units you travel horizontally. Equivalently, it means that however far you travel horizontally, you travel vertically by a fifth of that.

This is how the "slope" or gradient of a line is defined in math: by working out how far we go up vertically as a proportion of how far we go along horizontally. Expressed mathematically this says that the gradient is the following ratio, or fraction:

$$\text{gradient} = \frac{\text{vertical distance}}{\text{horizontal distance}}$$

Now in the triangle from our original unit circle we know that the vertical and horizontal distances are sine and cosine respectively:

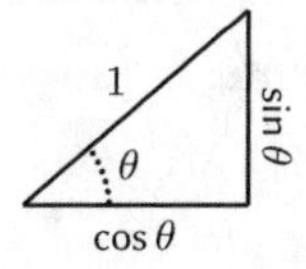

That means the gradient is $\frac{\sin\theta}{\cos\theta}$, and as the function tan is defined to be the gradient we duly get the formula:

$$\tan\theta = \frac{\sin\theta}{\cos\theta}$$

We can also do a quick "sanity check" and verify that if we scale this triangle up then the ratio won't change:

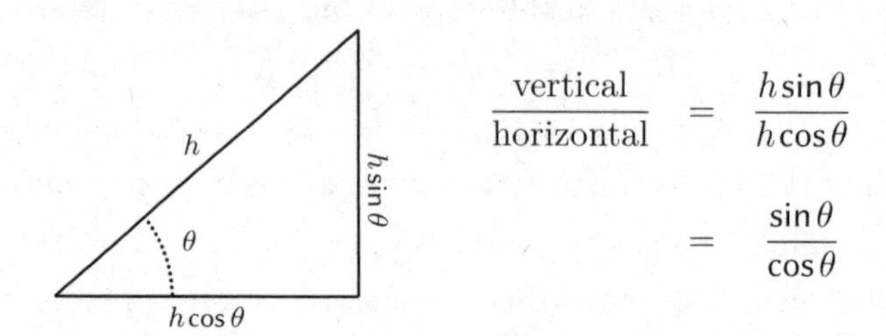

It might seem like you have to "remember" whether it's "vertical over horizontal" or the other way around, but I prefer *understanding* that the thing we're measuring is the rate at which the line goes up, which is a vertical distance. We then scale it by the horizontal distance as we're not measuring the *amount* we go up, but the *rate*, which means how far up we go relative to the horizontal distance.

Now it's true that if you're in school and burdened with the task of answering as many questions as possible in a short space of time on some sort of standardized test, you don't have time to go through that entire exploration to arrive at those formulas, and it might be beneficial to be able to produce them off the top of your head. But the problem there is with the exam system: the only reason to need to produce them at speed is because of these sorts of timed exams. Plus, I'm not sure those timed exams serve any purpose except finding some way to rank students in a hierarchy for the purpose of filtering them into different universities, jobs, or walks of life. It's a very poor justification indeed.

The upshot is that formulas are really explanations of things, and in this case they are explanations—and explorations—of the relationship between circular coordinates and square ones.

Circles and squares

Thinking about the relationship between circles and squares is also where the number π comes from. Thanks to Pi Day, π is now one of the most famous concepts in math, and it has happy delicious associations thanks to the fortunate pun with pie. Of course, that pun is rather Anglocentric, as I don't *think* there are many other languages in which the food sounds like the Greek letter. As an aside, I've just discovered that the Greek for pie (the food) is pita, which means that we Anglophones who think pita is a type of Greek bread have got the wrong end of the translational stick yet again.

Aside from the pi(e) pun being Anglocentric, the concept of Pi Day is also rather US-centric, because it depends on writing the date the US way around, with the month first, so that 3.14 represents March 14th. This, along with the arbitrariness of the pi–pie relationship, used to bug me about the concept of Pi Day, but I've come to realize that it's just a piece of fun and is one day a year when there's a general sense of fun around math, and there's no need for us to be so pedantic about the root of that fun. If we pour cold water on that fun then we're just cementing the impression that not only is math not fun, but it is actively *anti*-fun.

There's another objection to Pi Day that is even more pedantic (according to my definition of pedantry): some people argue that π is the "wrong" constant, and it "should be τ." This is the Greek letter *tau* and is used to represent the number 2π. We'll come back to that after thinking some more about what π really is.

It might seem like the point of the number π is to memorize as many digits of it as you can. One of the things I like less about Pi Day is the proliferation of competitions and "challenges." I don't like competitions in general (in $x + y$ I wrote all about them being too ingressive for me) and so I'm also a little uncomfortable with the proliferation of pie-baking *competitions* for Pi Day. I'm even more uncomfortable with competitions to recite as many digits of π as possible, because even aside from the competitive aspect, they focus our attention on the idea of rote memorization of numbers. There isn't really anything to understand about the digits of π—it's an irrational number and so part of the entire point of it is that the digits have no pattern.

So committing the digits to memory really has to be by rote, not by understanding.

I admit I quite enjoy making a joke about the fact that I know just two digits of π: 3.14. (The truth is that I actually know 3.14159.) Two digits is plenty for the level of accuracy I need in my life. It's not as if I'm doing some sort of precision engineering project where lives are at stake—for my purposes 3 would probably do just fine. That's because I only ever need π when I want to convert a recipe for a round cake into a square cake, or vice versa: it's about the relationship between circles and squares again.

A more precise way to phrase this is: If I have a recipe for a round cake, but I want to bake it in a square tin, how big a square is it going to make (assuming I want the depth of the cake to stay the same)? This comes down to approximating the area of a circle by a square, and is a conundrum that was considered by mathematicians in Babylon and ancient Egypt. Babylonian mathematics is very well-documented on clay tablets, and ancient Egyptian mathematics has been found recorded in a papyrus transcribed in the second millennium BC by a scribe named Ahmes, who announces that he has copied this from an older scroll. The papyrus is thus sometimes known as the Ahmes Papyrus, but unfortunately it is also known as the Rhind Papyrus, after a Scottish antiquarian, Alexander Henry Rhind, who bought it in Egypt in 1858; it was subsequently acquired by the British Museum. I hope that if any of my forgotten work is unearthed in the future and deemed worthwhile, it is named after me and not after someone who buys and sells it, even if naming it after the trader rather than the originator is the colonialist convention in a famous colonialist museum.

Another example is the controversial term "Elgin Marbles" for the Classical Greek marble sculptures made by the sculptor Phidias and his assistants for the Parthenon temple, and taken from there by the 7th Earl of Elgin in the early nineteenth century. I don't think we should name things after someone who just bought them, and we certainly shouldn't name them after someone who just took them from their rightful home, probably without any sort of permission.

Let us return to the subject of squares and circles. Ancient mathematicians weren't phrasing this question in terms of cake, but rather, they were stating it abstractly: Given a circle of a particular size, what size square has the same area? The question of the area of a circle requires a bit of a digression because it turns out to be rather difficult to define "area" where curves are involved.

The concept of area

We can *imagine* that finding the area of a curved shape is like pouring some liquid into the shape in a very thin sheet, and then pouring the same liquid into a thin square sheet and seeing how big the square is, but that is very far from rigorous. And it's very difficult to make it rigorous.

What we might do in elementary school is draw a shape on a square grid and then count the squares inside the shape. Then we need a scheme for dealing with partial squares: if there's a partial square that's at least half inside, we count it as one. If there's a partial square that's less than half inside the shape, we don't count it. Here's an example, with the squares I've counted shaded in:

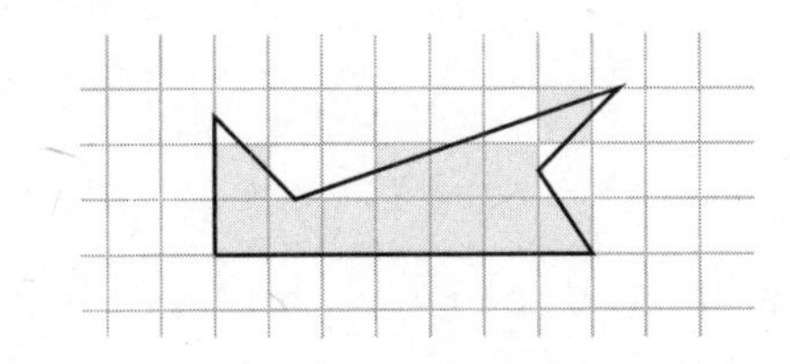

This is still not at all rigorous. Besides having to guess which squares are half in and which aren't (I wasn't really sure about the one I shaded at the top right), it's relying on the idea that the partial squares that are more inside will balance out those that are more outside when we add them up. It's not bad as a method of approximation, but it's definitely not a rigorous definition.

A more refined method involves dividing the shape up into triangles, and taking the areas of those. We work out how to find the area of a triangle by relating it to a rectangle: every triangle is half of a rectangle, and the area of the rectangle is pretty obvious. Or is it? Where does that formula even come from? Well it really comes from a leap of imagination from lining up grids of counting blocks, and also the idea that if we have a square whose sides have length 1, then it seems that we should declare that to have area 1. We can then build up rectangles of whole numbered sides, and then fractions, and make a leap of faith to irrational numbers.

So we've gradually built up the idea of area starting from squares, moving to grids, then rectangles, then triangles. This enables us to define, rigorously, the area of any shape made with a finite number of straight-line sides, because we can always divide one of those up into a finite number of triangles. But that in itself takes a little proving, and there's also the issue that there are many ways to do it, which raises the question of whether they'll all give the same answer. For example, it's not very evident why the following ways of chopping up the shape into triangles *will* produce the same answer for the total area, even though it might seem clear that they *should*.

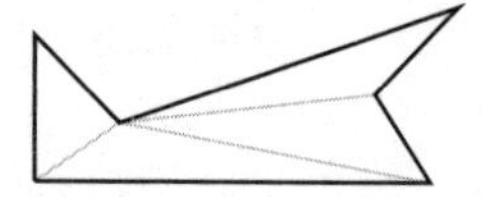
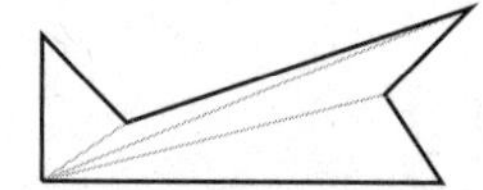
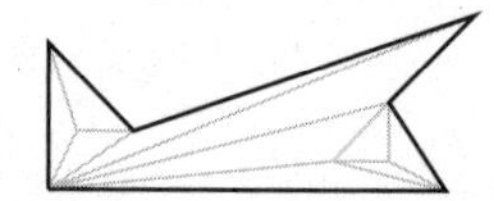

Again it's a scenario where people who think it's "obvious" might seem like better mathematicians, but really if you're baffled by it then you might be thinking more like a mathematician deep down.

The next step in this journey of exploring the concept of area is to wonder about curved edges. After all, nothing in nature has perfectly straight edges. This exploration is a key example of building up ideas from previous ideas. Can we take something with curved edges and relate it to what we already understand about shapes with straight edges? If something has curved edges we can still approximate it with triangles, but the triangles are always going to be a little bit off. However, this is a method that was used all the way back in around 250 BC by Archimedes of Syracuse. He approximated circles by regular polygons, that is, shapes with equal length straight sides and equal angles. The more sides you use, the closer your approximation will be, as we illustrated in Chapter 4 with squares compared with octagons.

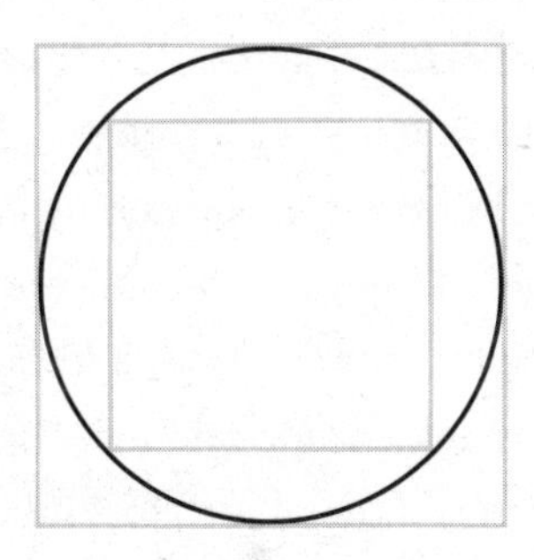

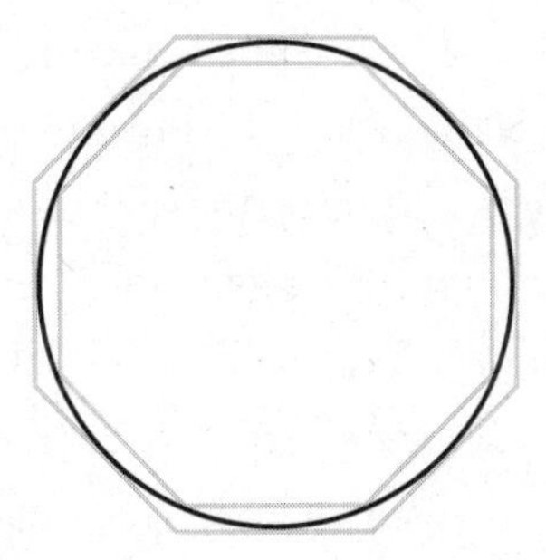

This helps us get closer and closer, if we use more and more sides, but it's not rigorous. The area enclosed by a curved line needs a whole lot of calculus in order to define it rigorously. It involves thinking about "infinitely small" triangles, and any time we're trying to think about infinitely small things, we need to invoke calculus. Well, there might be other ways of doing it but calculus is the most famous and arguably the most productive.

Actually, calculus is even needed to define the *length* of a curved line. The intuition behind the length of a curve is not so difficult—we can imagine running a piece of string along the curve, and then pulling it straight to measure it. But that's not rigorous.

Perhaps you feel that's good enough, and this is where the idea of development and building arguments in math comes in. The string method is perfectly good enough for everyday uses. If I'm lining a round cake tin with parchment paper, I don't calculate the circumference via π and the radius—I just wrap the parchment around the outside and cut it, allowing for some overlap. There's nothing in my daily life where I'm likely to need to know a curved length more accurately than that.

But when math piles structures on structures, and arguments on arguments, it needs things to be logically correct, not just more-or-less accurate. The logical correctness enables us to develop more math, and also to develop much more intricate applications such as in very complex engineering constructions. We already questioned the value of that type of "development" in the previous chapter. Besides that, there's the desire for sheer understanding, not just experimental data: using a piece of string gets us an answer, but what is really going on?

I am reminded of when I can't find something and am going around the house frantically looking for it. I don't like doing that; it feels like proceeding by experiment instead of by logic. I prefer thinking inside my head and working out what I last used the object for, and then deducing where it must be. I appreciate Agatha Christie's detective Hercule Poirot, who believes in solving crimes by thinking inside his brain and understanding *why* the crime was committed, rather than by scrabbling around on his hands and knees looking for physical evidence, like the other detectives he scorns.

If you feel no urge to *understand* lengths of curved lines, and also you're not interested in building up more complicated theories or applications, then it's true you may be left cold by the story of how calculus addresses this. That said, there's a meme that I've seen going around that people find quite amusing, which looks a bit like this:

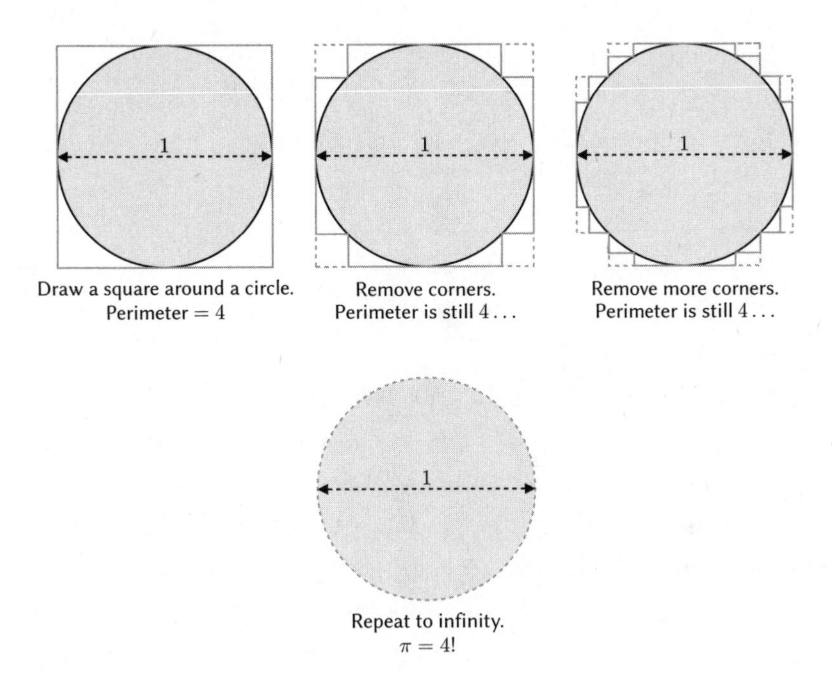

Draw a square around a circle.
Perimeter $= 4$

Remove corners.
Perimeter is still $4\ldots$

Remove more corners.
Perimeter is still $4\ldots$

Repeat to infinity.
$\pi = 4!$

Perhaps people share this because it appears to be breaking math. Many people who like math are drawn to try and explain what is wrong with the reasoning in this meme, but the deep mathematical answer is: it's not wrong, it's just working in a different context. Because as we've mentioned before, the concept of length is contextual. This means that π is also arguably contextual: it's not "just" a number, it's a relationship. And it's a relationship that highlights the different ways in which we can approach definitions and theorems.

What is π?

If we think π just "is" a number ($3.14\cdots$ something) then we have "facts" to remember like that the circumference of a circle is $2\pi r$ and the area is πr^2. I think this is a rather mundane approach to circles that hides something more wonderful, relating back to the principle of scaling shapes: that if we scale a shape in proportion, all the lengths get multiplied by the same amount. That means that the relationships between lengths in the shape stay the same. If we draw a rectangle where the long edge is twice the short edge, then no matter how we scale it in proportion, the long edge will still be twice the short edge:

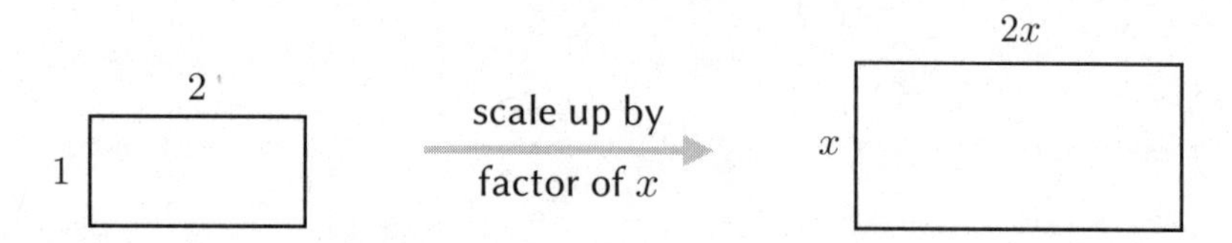

The idea is that this ought to be true for circles as well. We don't really know how to define the length around the outside of a circle (its circumference) rigorously, but whatever it is, it should have a constant relationship with the distance across the middle (called the diameter). No matter how we scale the circle, that proportion will stay the same. This is a wonderful, deep, fundamental fact, and honestly I don't really know where it comes from. Perhaps it's a mystery of the natural universe. A law of proportion? A truth about humanity? In some sense it's "obvious" but in another sense it's not obvious at all, and I like to marvel at it. It's funny how "obvious" can mean "it's so clearly true that I can't explain it." I recommend sitting and pondering that for a second. It seems that "obvious" means "inexplicable." That's extraordinary.

Anyway, the ratio of a circle's circumference to its diameter is thus the same no matter how big the circle is, which means it's a fixed number, a constant. Here's a case where using letters instead of numbers really comes into its own: we don't have to know what that number is, as a number, but we can still refer to it, as long as we give it a name. It's a bit like the fact that I don't know what the temperature outside is at the moment, but I can still refer to it as "the temperature outside" (whatever it is). So I can refer to this ratio, and mathematicians have picked the Greek letter π for it. After deciding to call it that, they *then* set about trying to pin down quite what number it is, by methods like polygon approximation. So what we're saying here is that the *definition* of π is

$$\pi = \frac{\text{circumference}}{\text{diameter}}$$

It just so happens that once we've pinned down what number π is, we can then rearrange this to find the circumference starting from the diameter, because we then get

$$\text{circumference} = \pi \times \text{diameter}$$

or $2\pi r$ since the diameter is twice the radius.

Now here's the subtlety: the ratio depends on what context we're in, because it's a ratio of *lengths*, and so it depends on what type of length we're talking about. And now we have two issues if we want to pin down what number π actually is: we need to know what type of length we're talking about, and we also need to know how to take the length of a curve.

If we go back to the taxicab world, in which we can only travel on a right-angled grid system, we can see how this context comes in. Remember, in that world, we can't go along diagonals, only along the "streets" that are on the grid system. What is a "circle" in this context?

Well, what is a circle at all?

When circles don't look circular

You might think a circle is a shape that looks like this:

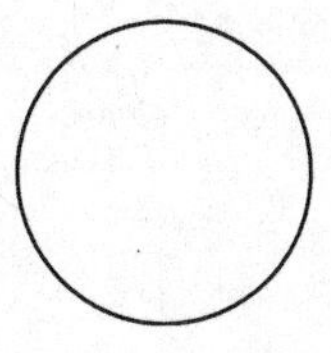

But what *is* that? How could we describe it to someone on the phone, or how could we explain the *concept* to someone?

The clue is in how we draw a circle with a pair of compasses (if anyone does that anymore, rather than selecting a circle function on a drawing app). You open the compasses to a fixed length, place the pointy end on the page in a fixed spot, and then move the drawing end around on the page. Because the distance from the pointy end is fixed, what you end up doing is finding all the points on the page that are the same distance away from the pointy end, which then ends up being the center of the circle. The fixed distance is the radius.

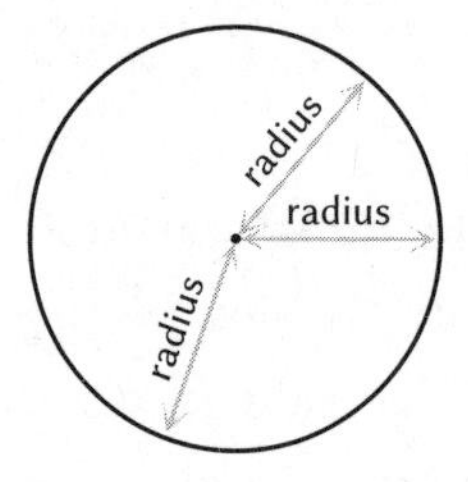

So a circle, really, deep down, is all the points that are a fixed distance away from a chosen center. This definition might sound remote and abstract, but as usual the abstraction has a point: we can now apply the idea more broadly. We can apply it in higher-dimensional situations, which gives us the concept of spheres and higher-dimensional spheres. And we can look for circles in other worlds where distance is measured differently, like in the taxicab world.

We could try and find all the points that are, say, four blocks away from this center point, which I have marked A:

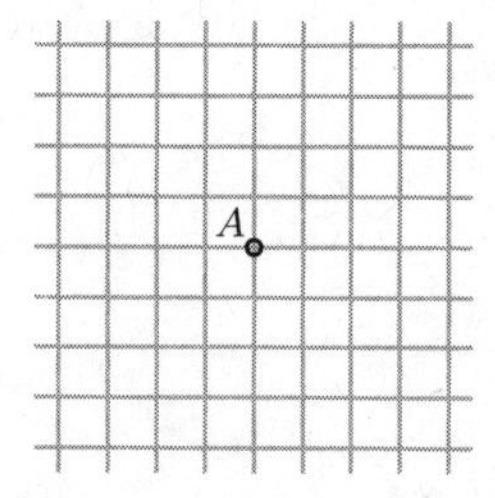

The most obvious places are directly north, south, east or west by four blocks. But we could also go two blocks east and then two blocks north, making a total of four blocks. We could also go three blocks east and then one block north. You can try zigzagging and going one block before turning, and one block, and one block, and one block, and that will turn out to be the same as just going two and then two. Remember not to go back on yourself, because then you're not really traveling the full distance.

If we mark all the points that are exactly four blocks away from the center point A we get this pattern:

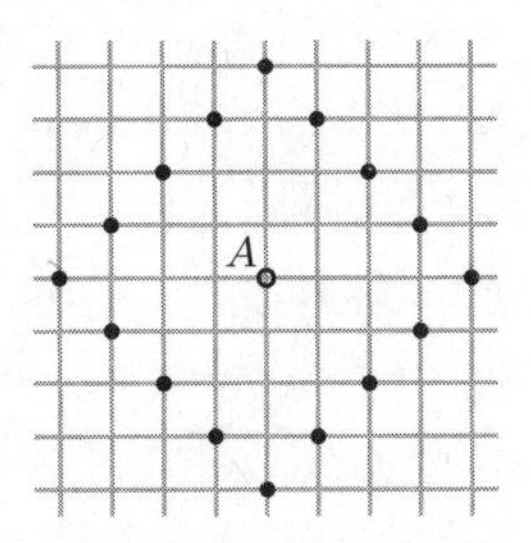

This is a "circle" in the taxicab world. Not only does it not look very circular, it's not even joined up by lines. It might be tempting to "join the dots" by drawing diagonal lines, but remember you can't actually go along diagonals in this world. It might also be tempting to then join them up with a step shape. You could go along those lines in this world, but the points in between would no longer be exactly four blocks from the center, so they're not actually part of this "circle."

We can now work out what "π" is in this world, because π is the ratio of the circumference to the diameter. But remember, we have to measure the distance *in this world*. To measure the distance around the outside we need to take the shortest distance between all the points, which means measuring the lines marked here:

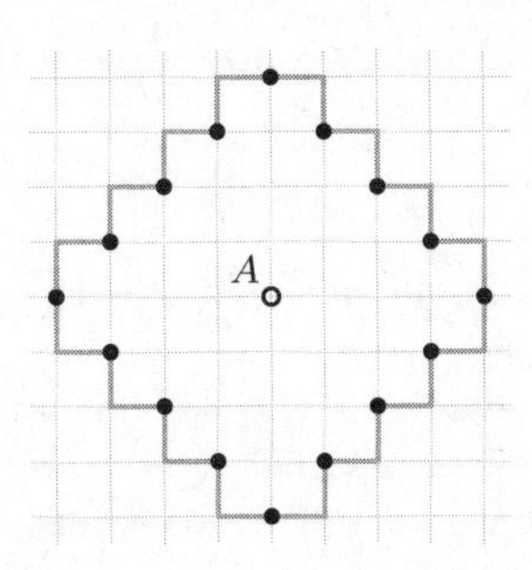

so we get a total of 32 blocks. The diameter is the distance across the middle, which is 8. So π is $\frac{32}{8}$ which is 4.

In the taxicab world, π is 4, just like in the meme.†

So π is not a fixed number, it's a ratio that will turn out differently in different contexts. Of course, we could just go ahead and declare that π is the number that we get from circles in Euclidean space (the "normal" distance we're used to), but I like focusing on context so I prefer thinking of it as contextual, personally. Also I like the idea of reciting all the digits of π in Pi Day contests by just saying "4."

But then again I've also decided that along with Pi Day we could have "i Day," for the imaginary number i, and that this should be the 29th of February because i is periodic: it goes around in cycles of 4 because i times i is -1, so i^3 is $-i$ and i^4 is 1. This means if you keep multiplying it goes i, -1, $-i$, 1 around and around and around, getting back to i every four, just like leap years. And I've decided that i has no digits so we've already finished reciting all of them.

I like the idea of i Day as a little pushback against the idea of π as a fixed number to be memorized. The number π is a fundamental constant, but it's a fundamental constant *relative* to a particular context, and in each context, the concept of π tells us something about the relationship between squares and circles in that context.

That's what the meme about π is really saying to me, although I suspect that whoever first made the meme didn't even realize that there was so much deep mathematics behind it.

I'd finally like to come back to that other constant that some people really believe in, called τ, the Greek letter *tau*. The idea is that the 2 in the formula $2\pi r$ is perhaps annoying and it would be better if

† I should probably point out that the meme is describing a slightly different situation as the thing drawn in the meme is not a circle in the taxicab world, and looks like it doesn't even exist in the taxicab world. However, it does exist in a version of the taxicab world that has streets infinitely close together, and in that case the circumference of that shape really is 4. It remains true that π is 4, it's just that this is not a justification of it.

we used 2π as a fundamental constant instead. This is the number that gets called τ. All this is saying is that we could take the ratio of the circumference to the *radius* instead of the ratio of the circumference to the diameter, and it means that the formula for the circumference is then τr (*tau* times the radius), which is a tidier relationship between circumference and radius. This is all very well but then we make a mess of the relationship between area and radius, because the area is usually πr^2, but if we express it in terms of $\tau = 2\pi$ instead we'll get $\frac{\tau}{2} r^2$.

What really bothers me about τ is that some people use it as a way of claiming superiority over other people, as if they are somehow privy to a deeper secret of the universe because normal people use π but *special* people know about τ.

Unfortunately there are plenty of math memes that do this as well, and in doing so they attract millions of likes, comments and shares. Periodically another one goes around that is just some test of applying the "order of operations," that is, the standard order in which we do +, −, ×, ÷ and so on. This order is remembered by various different but equally silly-sounding mnemonics in different parts of the world—differing even between English-speaking countries.

Mnemonics

Bodmas, bedmas, podmas, pedmas, pemdas. I'm not even sure what these stand for, but they're all supposed to be ways of remembering the order in which we do operations in math. In PEMDAS, P is for "parentheses" (corresponding to B for "brackets" in the UK), E is for "exponent," then multiply, divide, add, subtract. Including "divide" and "subtract" is already a bit redundant because dividing is really the same as multiplying (by the inverse) and subtracting is the same as adding (the inverse).

I just googled "Pemdas meme" and the first thing that came up was exactly the one I had in my head, looking something like this:

WHAT IS THE ANSWER?

$7+7\div 7+7\times 7-7$

unfortunately most will get this

WRONG!

That last line is a classic hook to try and draw in the types of people who think that they are smarter than most people, and who think that unlike everyone else they can definitely get this right. Comments on posts like this tend to dissolve almost immediately into people giving different answers and telling each other how stupid they are that they can't even do bodmas/pemdas or whatever. I dislike these memes for many reasons. The two main reasons are first that these memes are there mainly to give some people a chance to assert their superiority over others, and secondly that they draw attention to one of the most pointless and boring parts of math that has nothing to do with anything that mathematicians do or think about.

I have several Public Service Announcements to make at this point. First of all: mathematicians do not sit around adding and multiplying numbers! Secondly, and perhaps more devastatingly: mathematicians do not care about the "order of operations." Well, perhaps I shouldn't speak for all mathematicians, but I've never met one who does care about the order of operations. The thing is that the "order of operations" isn't really math: it's just a convenient notational convention. We could pick a different notational convention and it wouldn't make any difference to the actual math. I could decide that we're actually going to do addition before multiplication. We still need parentheses (or at least some convention) to override the usual orders, so we can't *entirely* reverse from PEMDAS to SADMEP, but we could go from EMDAS to SADME. In that case we'd get the following translations:

EMDAS world		SADME world
$2 \times 4 + 5 = 13$	$\longleftrightarrow$	$(2 \times 4) + 5 = 13$
$2 \times (4 + 5) = 18$	$\longleftrightarrow$	$2 \times 4 + 5 = 18$

See, it's not really that earth-shattering. The reason we do it the way we do is, in my head anyway, because we typically stop bothering to write the multiplication sign, especially when we're dealing with letters. So we write $2x$ rather than $2 \times x$ and then it makes visual sense to have things physically close together when they're a connected unit that goes together. So, visually, this expression

$$2a + 3b$$

puts the 2 with the a and the 3 with the b. The aim here is to use notation that aligns with our visual intuition. However, if we insert × signs then that helpful aspect is lost:

$$2 \times a + 3 \times b$$

I absolutely can't imagine any mathematician actually writing that string of symbols down, because it looks so confusing. Thus, I also don't think that any mathematician would ever write the expression in the meme, even if they were trying to do that calculation (which is unlikely). Not only do mathematicians (in my experience) rarely write × signs, they even more rarely write ÷ signs, preferring the much more visually compelling fraction notation. Then you don't even have to worry about an order of operations because something like

$$\frac{2}{5} + 7$$

makes it visually quite clear that the 2 and 5 go together, and the 7 is separate.

As for that expression in the meme, I would write it like this:

$$7 \ + \ \frac{7}{7} \ + \ 7 \cdot 7 \ - \ 7$$

That dot between two of the 7s is what we sometimes write instead of a × sign between two numbers, because we can't just write them next to each other like $7a$ because it would look like 77.

If I really had to use times and divide signs, I would insert parentheses and spacing to make it clear what was meant:

$$7 \ + \ (7 \div 7) \ + \ (7 \times 7) \ - \ 7$$

It's an abomination to me to write that expression without parentheses. And this is not a question of math, it's just mathematical orthography.

A section on mnemonics in which I'm ranting about my least favorite ones would be incomplete without a mention of FOIL. I already mentioned this mnemonic in Chapter 4—it is supposed to help us multiply out pairs of parentheses such as:

$$(2x + 3)(4x + 1)$$

The idea is that we multiply the First pair, then the Outer pair, then the Inner pair and then the Last pair. There are many mathematical issues with this beyond the fact that it enables us to bypass understanding. It actually *limits* understanding, by neglecting the fact that we really don't have to do it in that order: if addition is commutative then we could do F, O, I, L in any order, but even if we don't know that it's commutative we could multiply these parentheses out and get FIOL instead of FOIL. It's because the way of multiplying them out comes from the distributive law for multiplication over addition. There are actually two laws:

$$a(b + c) \ = \ ab + ac$$

and

$$(a + b)c = ac + bc$$

It might help to think of this like the "grid" method for multiplication, in which case the two forms of distributivity are represented by these pictures:

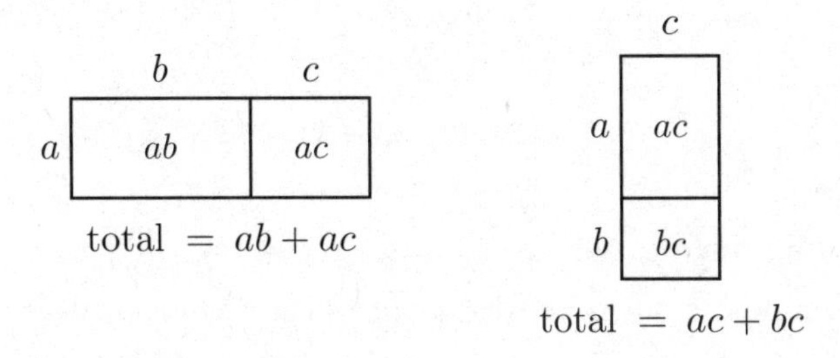

One of them is like stacking boxes horizontally, and the other is like stacking them vertically.

Now there's something rather profound here, that FOIL doesn't touch on at all: in order to multiply out something with two things in each parentheses, we have to use both these versions of the distributive law. Say if we're doing $(a + b)(c + d)$ it corresponds to this grid:

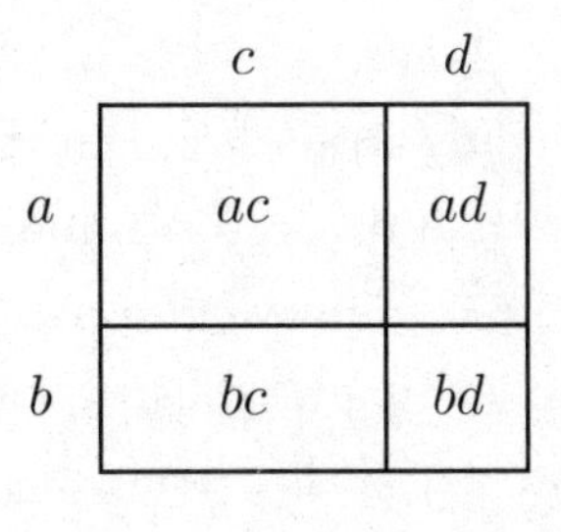

$$\text{total} = ac + ad + bc + bd$$

To get to the total we have to understand *both* the vertical stacking and the horizontal stacking of boxes, and decide which to do first.

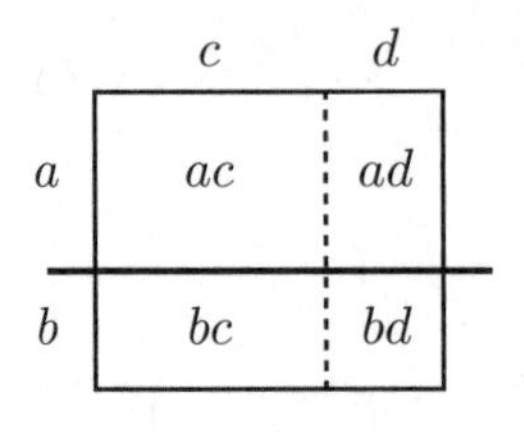

$$\text{total} = ac + ad + bc + bd$$

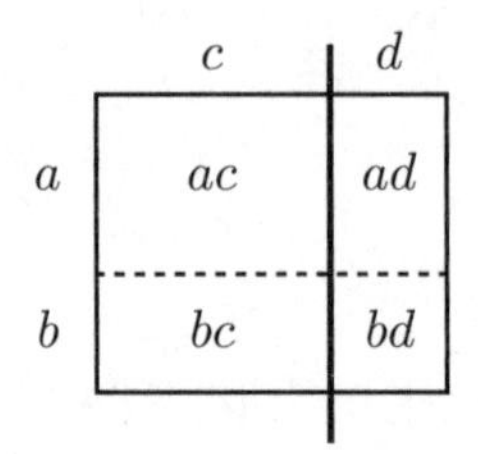

$$\text{total} = ac + bc + ad + bd$$

If we write this out in algebra, the first one consists of treating $(c + d)$ as a single unit first, giving

$$a(c + d) + b(c + d)$$

and then multiplying that out. This one is FOIL.

The second one starts by treating $(a + b)$ as a single unit, giving

$$(a + b)c + (a + b)d$$

and then multiplying that out. This one is FIOL. Now let's look at those two results. They are not quite the same—the middle terms are the opposite way around from one another, just like the middle two letters are swapped to turn the word FOIL into the word FIOL. If both distributive laws are true, then FOIL and FIOL have to give the same answer and it's a short step from there to deducing that addition *has* to be commutative. It has no choice.

I went through that slightly arcane argument for two reasons. First to show that fixing the order as FOIL is very limiting—the fact that FIOL and FOIL will both give the same answer does in a way mean that you only need to remember one, but also the fact that they're the same is demonstrating some quite deep math. The other reason is to emphasize the shortcomings of the FOIL method over the grid method.

The grid gives us a visual way of not just *doing* the multiplication, but feeling *why* it produces those four combinations of products. It's also a way of making four sandwiches at once: you put two fillings on one piece of bread, and two fillings on another, and then close them up perpendicular to one another. This produces four combinations of fillings for the sandwich, and also some vivid imagery.

The visual connections we make with abstract concepts are very important and are an entire part of mathematics in themselves. This is the subject of the next chapter.

CHAPTER 7

PICTURES

Why does 2 + 4 = 4 + 2?

This might seem innocuous and obvious, something that we all learned ages ago, something we can just see is true because both sides equal 6. Perhaps, given what we've already seen in earlier chapters, it might seem that we should immediately turn this question into "*Where* does 2 + 4 = 4 + 2?" This is a good thought, but I want to try a different approach this time and point out that actually those things are not *really* the same. The left-hand side means we're taking 2 things and then 4 more, and the right-hand side means we're taking 4 things and then taking 2 more. It's worth observing for a second how unobvious it is that those should turn out to give the same totals.

I'm going to explore how we can tell those give the same totals in principle, that is, without needing to know what the totals are. We can do it the way children first do it, using blocks or other objects, or by drawing pictures of objects. Using pictures might sound like a "childish" way to do math, but in this chapter I'm going to explore the profundity of pictures in math. I prefer to think of pictures as being "child-like" in the best way. "Childish" is a judgmental way of describing things, as if to say that it's something only underdeveloped humans do, and that they should really grow up and stop doing it. There are many aspects of being "child-like" that are wonderful and helpful, but which are unfortunately too often squashed out by the pressures of society or adult responsibilities. There's unbridled curiosity, openness to new ideas, fearless use of imagination, and a general comfort level with not understanding things. Children know that there are tons of things about the adult world they can't understand yet, and they're used to it. I'm sure this is one of the reasons they're less afraid of math, despite not understanding it. Whereas when adults don't understand it they think they're bad at it (or they remember being told they're bad at it).

Drawing pictures is another thing that researchers in abstract math do more similarly to children than might be expected. In my field of research, category theory, there are pictures everywhere, instead of equations or lines of algebra. It's called "abstract algebra" but actually most of the reasoning happens using pictures (more formally referred to as "diagrams") and the use of pictures is one of the things that has made it particularly productive as a new field of math. There is even a type of mathematical diagram whose technical name is *dessin d'enfant*, French for "child's drawing," coined by the great French mathematician Alexandre Grothendieck in 1984.

When I was first commissioned to make mathematical art by Hotel EMC2 in Chicago in 2016, I did not think of myself as a visual artist. Then it dawned on me that I draw so many pictures as part of my research that I am, in that sense, an abstract visual artist. So this chapter is about what pictures do for us in math—they're not just visual aids, but they can become part of the actual math itself.

I'm going to use some pictures to help us see why $2 + 4 = 4 + 2$, seeking a deeper level of understanding than just saying "because both sides equal 6."

"All equations are lies"

It often takes children a while to be convinced that $2 + 4$ really does equal $4 + 2$, and that's fair and valid, because there is an important sense in which it's not really true. If a small child is still only doing addition by counting with their fingers, then if you ask them what 4 plus 2 is it's quite easy: they put 4 in their head, and "count on" by 2 with the help of two fingers, and get 6 quite quickly. However, if you ask them to do 2 plus 4, they'll put 2 in their heads, and then attempt to count on by 4 using four fingers. For a child in that phase, four might be quite a lot to have to do on their fingers, and so it will take longer, be more arduous, and possibly turn out wrong.

So for this child, 4 + 2 is really not the same as 2 + 4: the second addition is a lot harder than the first one. This is why, instead of asking why these are the same, or even where they're the same, I want to focus on *in what sense* they are the same.

In this case, the two sides are the same in the sense that they should produce the same answer, even though the processes involved are different. And this is really the point of the equation: one side is harder than the other to process, and so we really benefit from knowing that the two sides give the same total. It means we can use the easier side to understand the harder side. This is true of all equations that we use in math.

The equation we're talking about is an instance of the *commutativity* of addition of numbers, which says that it doesn't matter in what order we add things up. Another basic principle about addition of numbers is that it doesn't matter how we group them together, which we usually denote with parentheses—that's *associativity*. For example, if we're thinking about addition, the following is true:

$$(8 + 5) + 5 = 8 + (5 + 5)$$

Personally I find the right-hand side much easier to think about than the left. The parentheses tell me to do 5 + 5 first, and that's easy: I know that 5 + 5 is 10 without having to engage any part of my brain, and I can then add 8 to 10 without having to engage my brain either.

By contrast, the left-hand side asks me to do 8 + 5 first, and that is much harder because it goes over 10. Now, it's true that it doesn't tax me *that* much, but it definitely involves a more intense level of engagement of my brain than 5 + 5. Rather than boasting about how easy we find things, I think it's much more interesting to closely observe which things are harder than other things, especially as those tiny amounts of extra cognitive load can quickly pile up and exhaust us.

Incidentally, this is one reason that the memorization of "facts" is sometimes urged in math, the idea being that if you memorize these things then you can produce the answers without causing that cognitive load, and then you can keep going faster. I personally find I can reduce the cognitive load by internalizing rather than memorizing. I have never "memorized" the fact that 5 + 5 is 10; it's just a part of me, in this case rather literally, on my hands. It's important to note that while memorizing something might help reduce later cognitive load, it might also cause increased disengagement from understanding, and that the trade-off is often not worth it.

So anyway, the point of the equation 2 + 4 = 4 + 2 is that the two sides are different in some way, that one side is easier than the other, and so we can use the easier side for the process, knowing that the totals will turn out to be the same.

In fact, this is really the point of *all* equations: they're about finding two things that are different in one sense but the same in another sense. This means we can use the sense in which they're the same to pivot between the senses in which they're different, and thereby increase our understanding and move it further from where it was. We tend to focus on an equation telling us two things are the same, but this is crucially tied up with the two sides also having a sense in which they're different—so they're not really the same at all.

The only equation in which both sides are really truly the same is this:

$$x = x$$

and anything of that form. This equation really does have both sides being the same—and as a result it is completely useless. We don't gain any knowledge of anything by stating that something equals itself.

I sometimes like to say "All equations are lies," which is a little bit clickbaity of me I admit. I might qualify it by saying "All equations are lies (or useless)." But the point is that the equation isn't exactly *lying* about the two sides being equal, it's just more nuanced than we give it credit for. (Or maybe we're the ones who are lying about the equation.) Equations are really saying that if we decide, for now, to focus on one particular aspect of these two situations, then they come out equivalent, even though overall the situations are different in some meaningful way. This is an important part of my research in higher-dimensional category theory, because when we're studying concepts that are more subtle than numbers, we also have more subtle ways in which they can be the same rather than just by being equal. So we have delicate choices about what we're going to count as the same in any given situation. This brings us back to our original question about $2 + 4 = 4 + 2$, and back to how we would ever explain this to a small child using blocks.

The profundity of counting with blocks

Doing arithmetic with blocks is usually considered something that only small children need to do, before they've learned how to do arithmetic "properly." They're supposed to grow up and stop using the blocks, and just do things in their head, or use various snazzy "strategies" that often baffle their parents.

Actually, doing arithmetic with blocks is very profound and gives us deep hints at subtle aspects of abstract math in higher dimensions. It might be surprising that higher dimensions come into things when we're just doing basic arithmetic, but it really does.

We've already seen a little bit of higher-dimensional thinking when we did multiplication in grids, and this really comes down to how we can think of multiplication as repeated addition. If we think about 3×2 as 3 sets of 2, we might take three pairs of blocks like this:

and then if we think about 2×3 we might take 2 triplets of blocks like this:

At this point it's not so visually obvious why those are the same. We could just count them, but that would only confirm that they are the same, without telling us *why*. And without seeing the general principle, it's hard to generalize it to other numbers.

We have already seen (in Chapter 3) that it is more illuminating to put the blocks in a grid as shown below.

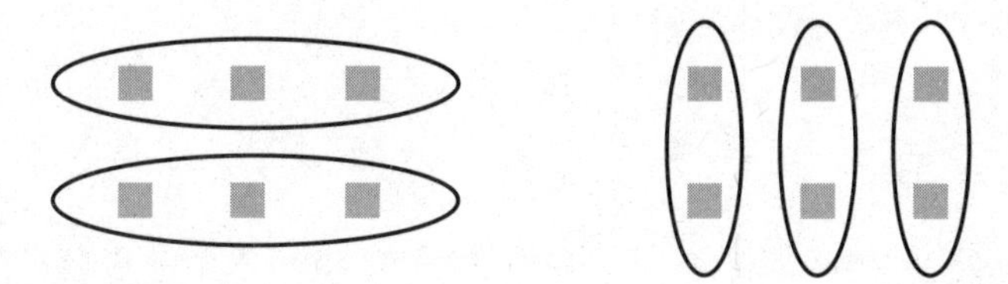

We can now see something deeper about *why* it's true, invoking more of our visual intuition. Crucially, we don't actually need to work out what the answer is in order to understand that the answer is the same in both cases.

This picture is depicting the fact that we can consider this grid of blocks as two rows of three, or as three columns of two. In a way, what we're really doing is using a rotation: we could think of rotating the table that the blocks are on, or walking around it to look at it from a different point of view. The number of blocks doesn't change just because we walk around it. Abstractly this is like when I said scaling a triangle up or down is like looking at it from closer up or from further away—we've changed our view of the triangle, but we haven't changed the triangle itself.

The logical structure corresponding to us rotating our view of the grid of blocks is:

$$3 \times 2 = 2 \times 3$$

However, note that this argument does involve us using two dimensions rather than just one. If these were beads on a single abacus rail we wouldn't be able to make that argument.

We even require an extra dimension if we think about commutativity for addition. If we imagine doing 4 + 2 with blocks, we might get this:

and if we do 2 + 4 we get this:

Now, we can understand that these are the same by walking around to the other side, or indeed by sliding the blocks around each other.

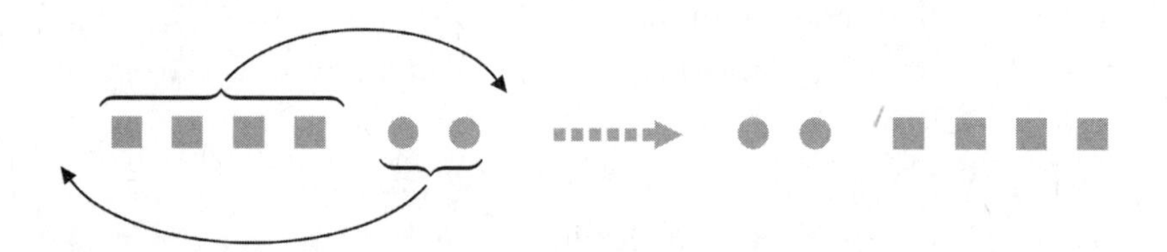

This is visually compelling, but again it's something we need two dimensions in order to do. By contrast, associativity can be done inside one dimension (such as on an abacus rail), just by sliding the circles along sideways as shown below. This picture depicts (4 + 2) + 3, before we slide them:

Afterward we'll get this, depicting 4 + (2 + 3):

This exploration has duly nudged us into asking *where* 2 + 4 = 4 + 2. It matters where we're doing it, because if we're in a one-dimensional world we won't be able to. Drawing the situation in pictures helped guide us to that.

The role of pictures

Sometimes pictures are just a visual aid to help us understand an abstraction, and sometimes pictures are a piece of formal notation for doing calculations. The latter happens a lot in category theory, where we are dealing with algebra that is more complex and so needs more help than just symbols written in a line.

I want to acknowledge that this is thus arguably ableist and excludes those who can't see. I try to include many vivid diagrams in all my books; however, I have had (very polite) messages from visually impaired people who have listened to audio versions of my books and appreciated them, but have been unable to refer to the diagrams in the included pdf. I don't yet have a solution to offer for that, and I'm sorry. For many other people, the visual representations in mathematics are extremely helpful and have enabled the subject to progress.

I will add that there are brilliant mathematicians who are blind, and many of them are actually working in geometry and topology dealing with shapes and relationships between shapes. Bernard Morin famously came up with a way to mathematically turn a sphere inside out (technically called "eversion"). This is so counterintuitive in the physical world that it is possible that *not* being able to see the visual physical world helped.†

I also want to acknowledge that relying on visual aids can be hard for those with hearing difficulties. I recently taught a student who was deaf, and who had sign-language interpreters attending classes with him. I realized it was impossible for him to watch my visual explanations while he needed to watch the signing at the same time. I ended up doing all the visual explanations twice but it was surely not ideal for him to have to see the signed explanation separately from my visuals.

I'd like to include a few examples of visual representations in my work in category theory. Here are some that are actual parts of rigorous arguments:

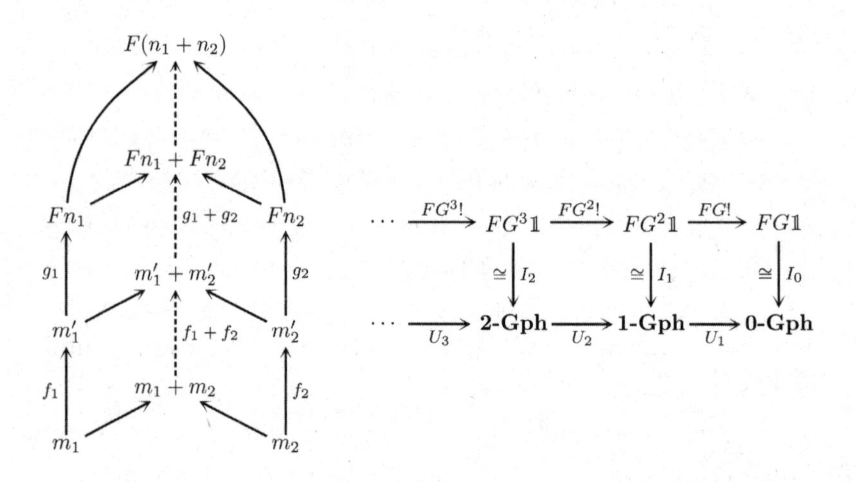

Here are some that are visual aids to help us with some intuition:

† There is an interesting account of the work of Morin and other mathematicians who are visually impaired, in "The World of Blind Mathematicians," *Notices of the American Mathematical Society* 49, no. 10 (November 2002), available via www.ams.org/notices/200210/comm-morin.pdf.

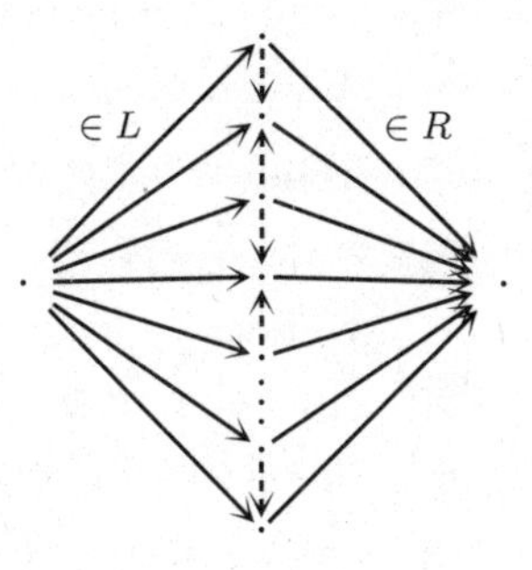

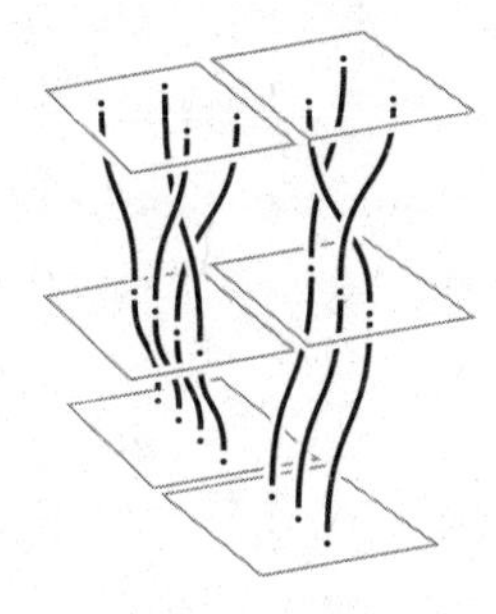

The key in both cases is to be able to understand how the logic corresponds to the visuals, and how the visuals correspond to the logic. This process isn't confined to high-level research mathematics: it is also the point of sketching graphs, something that tortures many students at school.

Graph sketching

Why is graph sketching so important? One complaint I've seen is that surely being able to sketch a graph well is more of an artistic skill than a mathematical one, so why do we have to do it in math?

My first answer here is that something being a question of art does not preclude it also being a question of mathematics. My deeper answer is that it depends what we mean by being able to sketch a graph "well." I do think that being able to sketch a graph "mathematically well" is different from being able to sketch it "artistically well" and that the two are both important but for different (but overlapping) reasons. In typical math education we're aiming to be able to sketch a graph mathematically well, but it's an oversight to ignore the artistic facet.

Incidentally, the word "oversight" is one of those odd words that can mean two opposite things, like "overlooking": it can mean seeing over something as in seeing past it, and missing it, but it can also mean seeing over something as in looking over it and seeing everything.

A project involving a large team of people needs good oversight, and if nobody is in charge of oversight that could be regarded as something of an oversight. Other words like that are impregnable, cleave, and resigned. They're called auto-antonyms, and I find them rather intriguing.

Anyway, graphs are often associated with painful and pointless math lessons where you have various formulas thrown at you without any claim that they represent something you're interested in, and you have to make pictures of them. The point has been completely obliterated: we've lost sight of why we draw graphs and how amazing it is as a process.

One of my earliest memories of math is a memory of my mother explaining to me that you can draw a graph of squaring. That is, you can take the process of squaring numbers and turn it into a picture like this: for every number on the horizontal axis you mark a point a certain distance above it, and the vertical distance you use is the original number squared.

So we start with this process:

$$\begin{array}{ccc} 1 & \longmapsto & 1 \\ 2 & \longmapsto & 4 \\ 3 & \longmapsto & 9 \\ 4 & \longmapsto & 16 \\ & \vdots & \end{array}$$

but we also include 0, and negative numbers and numbers that aren't whole numbers. It turns it into this picture:

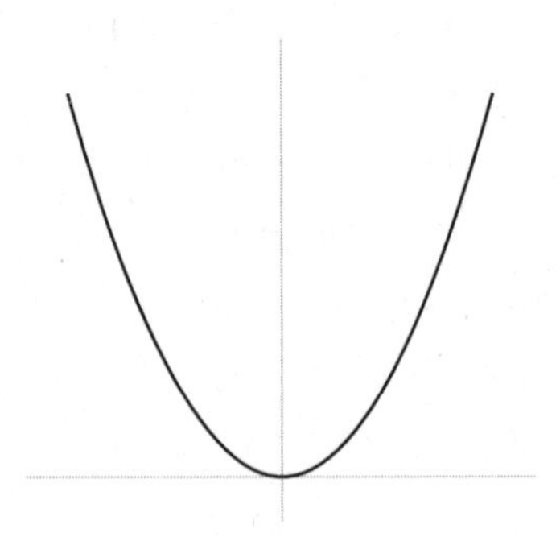

I was quite young and remember feeling my brain contorting out of my skull in all sorts of new directions to get my head around this shift from a mathematical process to a *picture*, and how amazing it is you can then *look* at a mathematical process. It now reminds me of what is called "intermodal translation," as explained to me by my dear friend Amaia Gabantxo, who is a literary translator, writer, poet, musician, and general creator. She teaches and practices translation not just between languages, but between different forms, perhaps from music into poetry, or poetry into dance, or food into dance. In a way, we could think of drawing graphs as a form of intermodal translation. The idea, as I see it, is to benefit from the strengths and intuitions in each of the different modes. In graph sketching the "mode" we start in is a mathematical process, rigorously and formally stated, usually using a formula. The formula is very good for reasoning with, and for building up logical arguments. We translate it to a picture, which is less formal or rigorous, and less good for building up logical argument, but it's much better for engaging other intuitions, perhaps more human intuitions, about shapes and trends.

For example, here's a list of numbers showing COVID-19 cases in New York City from March 2020:†

3/3	1	3/10	70	3/17	2452	3/24	4503
3/4	5	3/11	155	3/18	2971	3/25	4874
3/5	3	3/12	357	3/19	3707	3/26	5048
3/6	8	3/13	619	3/20	4007	3/27	5118
3/7	7	3/14	642	3/21	2637	3/28	3479
3/8	21	3/15	1032	3/22	2580	3/29	3563
3/9	75	3/16	2121	3/23	3570	3/30	5461

It's hard to form any sort of gut response to this list other than to see that the numbers are increasing. However, here they are as a graph; the

† Data obtained from https://github.com/nychealth/coronavirus-data.

one on the left is a bit wobbly, but the one on the right shows the running seven-day averages, which smooths out the daily fluctuations.

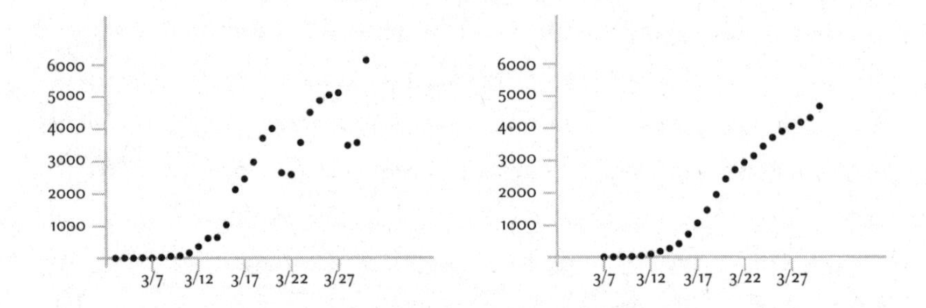

We can now invoke our visual intuition and see a very alarming increase, not just in the numbers, but in the steepness of the graph, until the slope stabilizes slightly around March 22.

Working with abstract concepts is hard because—well, because it's abstract. Translating it into a realm where we have more intuition can help. Translating abstract ideas into something visual so that we can use our visual intuition is a very fruitful part of mathematics. The important thing then is to know which abstract features correspond to which visual ones.

Translating features

Here are some of the visual features we might ask about when looking at a graph: Does it have corners? Holes? Is it going up or down? Does it turn around? Where does it go up or down faster? Does it shoot off to infinity anywhere? Does it go steadily to infinity at the ends? Does it flatten off? Does it fluctuate wildly?

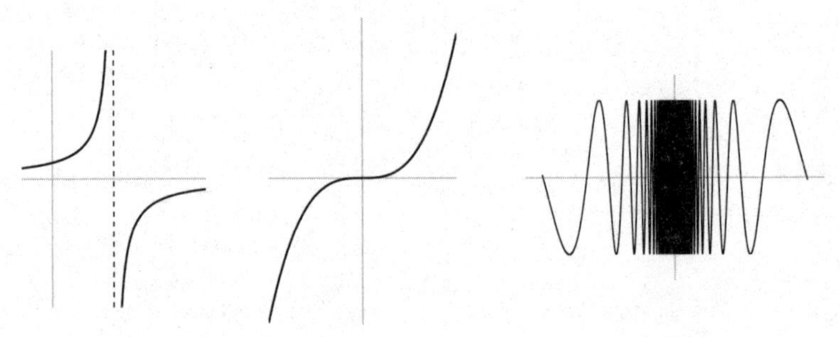

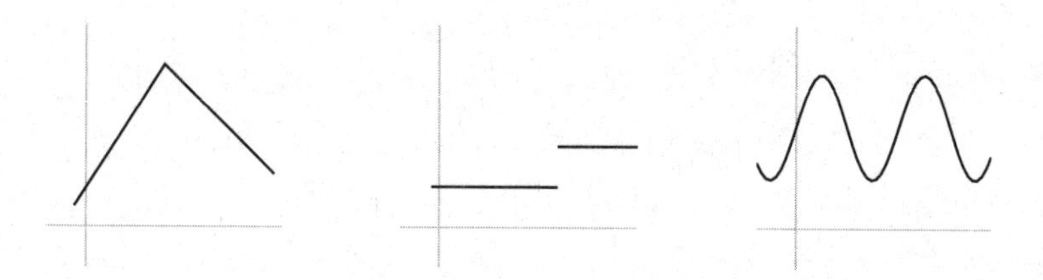

We then set about understanding why these things happen. The visual features are very vivid; what sort of logical structures do they correspond to? And then we can ask the question in reverse: If we start with an abstract situation, what sort of visual features does it produce? That's what sketching graphs is really about. It's a way of getting a better understanding of how these formulas work.

Sketching graphs well is not about producing beautiful curves, it's about clearly representing the key features that we're interested in. In a way, that's what art is about too—it's about deciding on some feature of the world that you want to draw the viewer's attention to, and then choosing a way of presenting it that really focuses our attention on that feature. Cubism and impressionism are choosing to draw our attention to different things, in different ways.

Now, in some situations, representing something "clearly" does mean representing it beautifully. It depends who your audience is. When I first started teaching at art school we were doing an activity involving the symmetry of an equilateral triangle, so I gave everyone a card and scissors and asked them to cut out a rough equilateral triangle. What I didn't take into account was that the students would be very concerned with trying to make their triangle perfectly equilateral. I hadn't yet told them exactly what we were going to be doing, and that it didn't matter for it to be perfectly equilateral, as we weren't building structures or fitting the triangles together in any way. It only mattered that they could regard it as equilateral in their head. That's still ambiguous because everyone has a different level of tolerance for what you can "pretend" is an equilateral triangle. I'm quite happy with these:

just like I'd be quite happy to draw these and declare them to be circles:

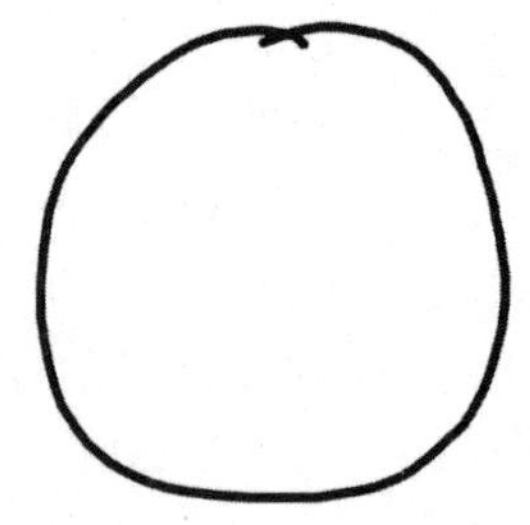 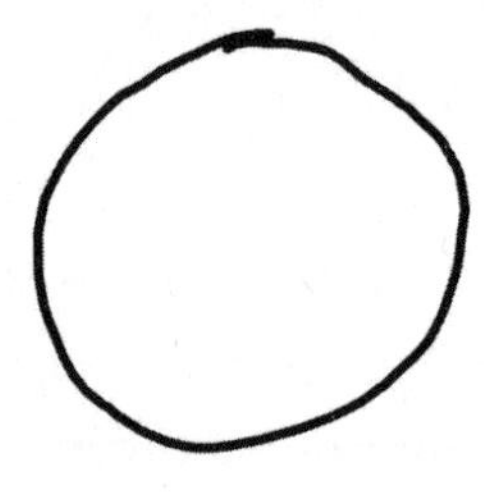

whereas a small child or someone with lower tolerances might be unable to think of those as circles. It's a bit like suspension of disbelief when watching actors, or when watching an opera singer who's supposed to be fourteen or consumptive. Those of us who are very enthralled and transported by music can be carried away to imagine practically anything while the music is happening, whereas others are less moved and unable (or unwilling) to believe that this person is supposed to be fourteen. A worse problem is when critics or directors think that someone playing a hero or heroine has to be thin, because otherwise the character is unbelievable; not only is that fat-shaming, but it's also really that critic's problem if they are unable to imagine people falling in love with a person who isn't thin. I digress.

For math students it can be baffling which features count as "important" and which don't, especially because that changes according to context and this is often not made clear. Educator Christopher Danielson points out that if you show children these two shapes:

many of them will say that the one on the left is a square and the one on the right is not—it's a diamond. But they're both the same square, just different ways up. We may teach them that those are the same shape, but then if they write their numeral seven like this:

they are told it's the wrong way around. See: for shapes we call them the same no matter which way up they are, but for numbers and letters it really matters which way up they are. The context matters.

We saw this in Chapter 1 when talking about different concepts of triangle in different contexts. In order to know which features are important to highlight, we have to know (or decide) what context we're in.

You might say: Why do we still bother sketching graphs by hand when computers can do it and very accurately depict *all* features? This is a good point, and when graphic calculators were first invented I loved just typing formulas in and watching it draw the graph for me. What a relief! Nowadays you can just type a formula into a search engine and it will draw a graph for you, and I actually quite enjoy doing that. From time to time I just type "sin (1/x)" and then "x sin (1/x)" into the search bar and marvel at the graphs that are instantly produced. I find them spectacular. Here they are:

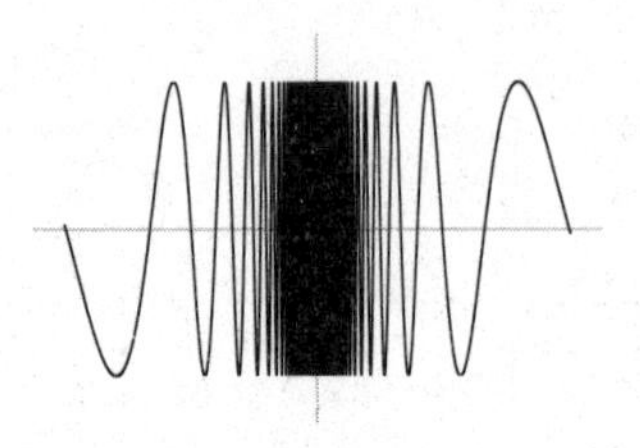

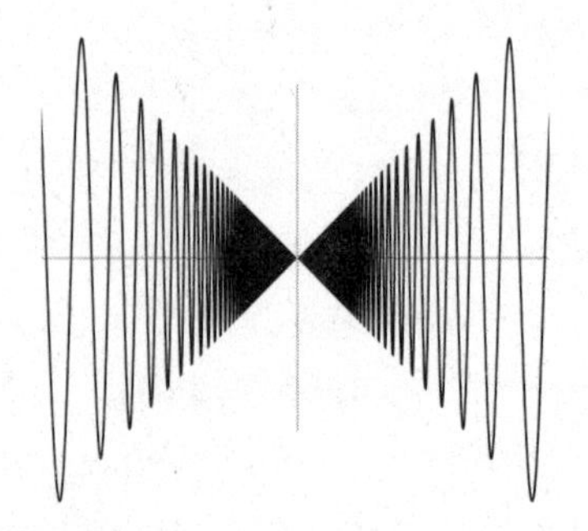

However, the point of sketching graphs isn't to produce a picture of the graph; the point is to *understand* what algebraic, logical features are producing which visual features. So even if you have a picture in front of you, it remains to understand *why* that is the picture of the graph. If the question "Why?" pops into your head when you look at a graph like that, then you are thinking like a mathematician. If your instinct is to run away then perhaps you've just been too traumatized by graphs in the past, and that is the fault of the education system. (Please note I'm not blaming individual teachers, but the system that constrains them.)

Representing math in pictures is profound and also baffling because it involves understanding or deciding what is relevant in the current situation, and only representing that. Suppressing other features can especially help people who are not so well versed in the situation and so don't know which features to ignore. Sometimes the "artistic" features are important as well, but sometimes mathematicians neglect that because they're so used to thinking abstractly.

This is a really important aspect of data visualization, and is something that Florence Nightingale understood well.

Florence Nightingale, mathematician

Florence Nightingale is also known as "The Lady with the Lamp" and is probably most widely remembered for being a wonderful nurse. The thing is that she was actually a brilliant mathematician and statistician as well. Arguably her great contribution was in doing rigorous quantitative analysis of how soldiers were dying during the Crimean War. The mortality rate before her intervention was as high as 40 percent, but her analysis led her to estimate that ten times more soldiers were dying from disease rather than from the actual war. She implemented the steps needed to dramatically reduce those deaths, including cleaning up the hospitals, and improving ventilation and sewage as well as the soldiers' diets. But moreover, not only did she do this analysis, she also understood how important it was to communicate it clearly and vividly to people in power who might not understand data as she did, and so she devised visually compelling ways to represent that data. What she came up with was a version of a pie chart, which she called a "coxcomb," but which now has the rather mundane name "polar area diagram." Here's an example. The diagram represents deaths month by month, with each month divided into deaths from battle wounds, deaths from disease, and deaths from other causes.†

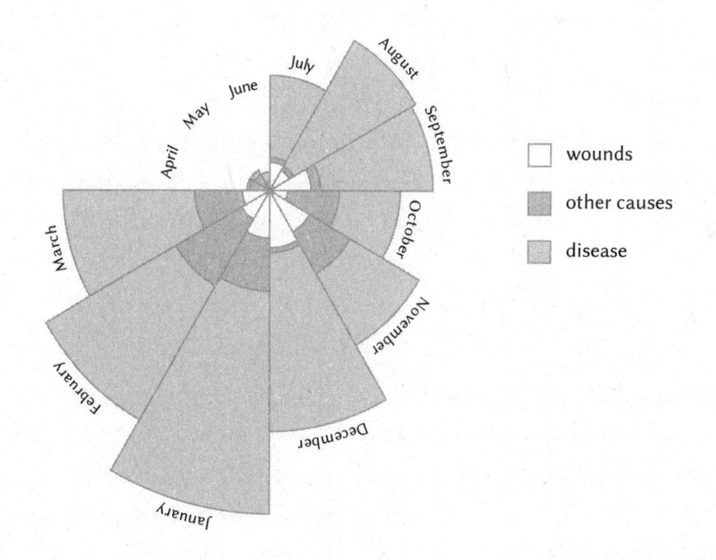

† I have sketched a part of Nightingale's original diagram entitled "Diagram of the causes of mortality in the army in the East," which was published in *Notes on Matters Affecting the Health, Efficiency, and Hospital Administration of the British Army* and sent to Queen Victoria in 1858. A full reproduction of the diagram can be seen at https://en.wikipedia.org/wiki/Florence_Nightingale#/media/File:Nightingale-mortality.jpg.

What matters here is the pattern of relative rates of mortality, not the sheer numbers themselves. This diagram makes it very visually vivid that deaths were not only dominated by disease, but moreover were highly seasonal. The actual numbers are represented by the areas of each piece of the diagram, which is quite complicated to calculate, but visually arresting. A pie chart is a simpler presentation, but again what matters is the proportions not the sheer numbers. Here's a pie chart representing uses of water in the US:†

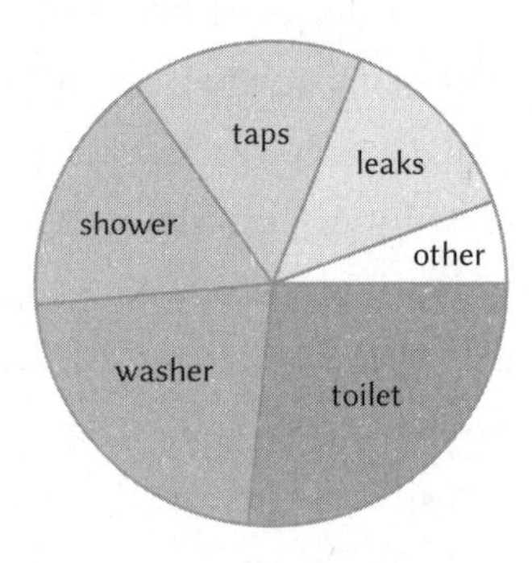

The numbers in this sort of pie chart are also represented by area, but the areas are easier to calculate as they're just proportions of a circle. The diagram is simpler because we care about even fewer aspects of the situation: we're not simultaneously looking at months of the year, just the relative proportions of the different types of water use.

This idea of only caring about one thing and not others under certain circumstances reminds me of the Chicago winter, where at a certain point it gets cold enough that I no longer worry at all about what I look like, I just do whatever it takes to remain warm, and possibly look mildly ridiculous in the process. Sometimes I wish I could achieve that level of freedom in the summer as well, but social pressure still gets the better of me.

† I found this data at www.statisticshowto.com/probability-and-statistics/descriptive-statistics/pie-chart/. It comes from a study by the American Water Works Association Research Foundation, "Residential End Uses of Water" from 1999. So it's rather old but I'm only using it to illustrate pie charts.

Pie charts and polar area graphs are about visual representation of data. However, in abstract math we are not just thinking about "facts" but also processes, so we need ways of visually representing processes. This is a central part of my research in category theory. One very vivid example of this is in how we study more nuanced aspects of commutativity.

Commutativity with nuance

The visual explanation of why $2 + 4 = 4 + 2$ helped us understand more deeply why it's true, but it also opened up more questions. We saw that, as long as we're not stuck in a one-dimensional world, we can justify this equation by sliding blocks around each other like this:

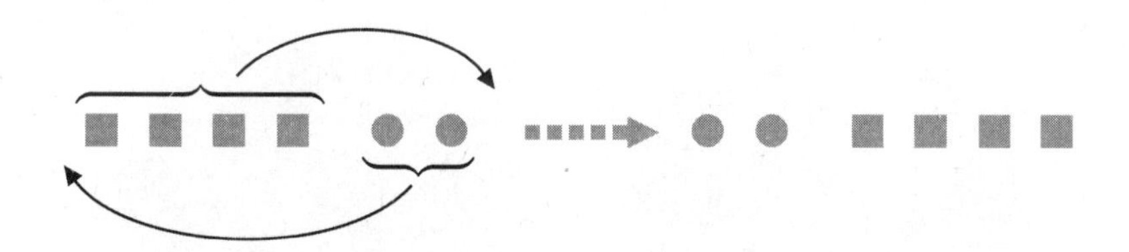

But there is another way to slide them around each other:

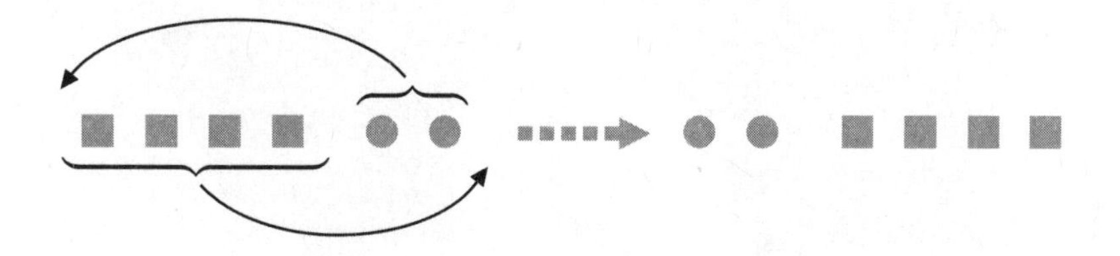

So there are two slightly different ways in which $4 + 2$ turns out to be the same as $2 + 4$. In abstract math we care more about the processes involved in getting to somewhere rather than just the destination. So maybe we care about which way around we did that sliding.

We might care just because we're interested, or it might actually make a difference. In the old English tradition of Maypole dancing, dancers hold on to long ribbons attached to a Maypole in the middle. The dancers dance around each other, and in doing so the ribbons are woven around the pole. I was really intrigued by this when I was little but didn't think of it as math until later. It really matters which way the dancers weave past each other, because it will create different patterns on the pole, and if they don't weave in and out then it won't weave any pattern at all.

That's a bit esoteric, but the idea also comes into play if you braid your hair, especially if you're doing a French braid. Here are two pictures of my hair where in the first one I've brought the strands from the outside *over* the strands in the middle to make a French braid, and in the second one I've brought them *under*, which is also known as a Swedish braid.

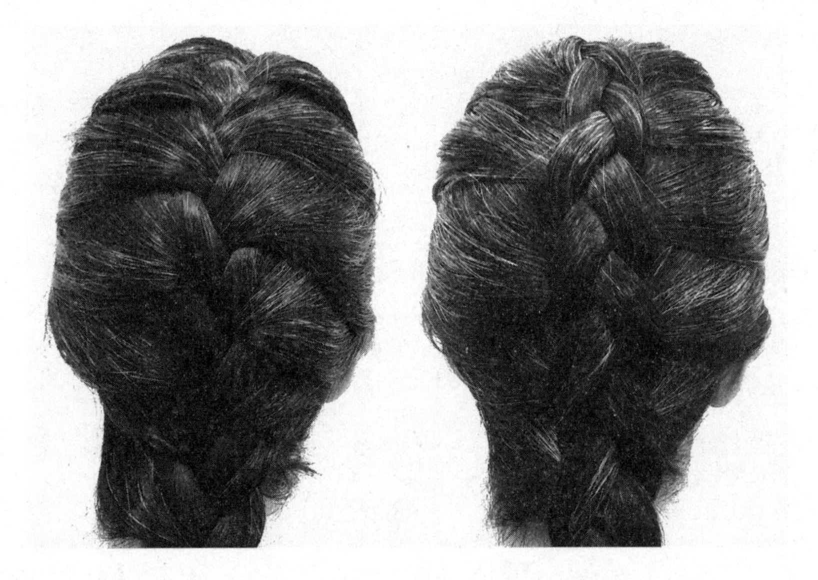

When I was little I was always intrigued by how those different styles could be made. I'm now meta-intrigued that this has connected into part of my research in higher-dimensional category theory.

The starting point of that part of my research is that we are often not just interested in whether or not something commutes, but we are interested in the *ways in which* it commutes. And if it doesn't *quite* commute, in what way does it fail? Does it "sort of" commute? These are much more subtle and deeper questions than just saying yes or no in answer to "Does it commute?"

Some of my favorite examples are in the kitchen: if you make mayonnaise you really need to start with egg yolks and then add the olive oil slowly. If you start with olive oil and try to add egg yolks then it won't work: it doesn't commute. It doesn't even slightly commute. It's important to understand when things do and don't commute in the kitchen so you know when it's crucial to add things in a particular order, and when it isn't. When I'm making chocolate mousse I find it important to add the egg yolks to the melted chocolate (slowly) rather than add the chocolate to the egg yolks, which is liable to make the chocolate seize. When making tiramisu I've tried adding the mascarpone to the egg yolk mixture but I've had better success adding the egg yolk mixture to the mascarpone, otherwise the mascarpone leaves lumps in the egg yolk mixture. I read somewhere that you could get rid of them by heating it gently, but all that did for me was make the whole thing runny, and it was very disappointing. (Someone is going to want to write to me now with their tips for how to make better tiramisu, but I'm really not asking for advice.)

In abstract math we seek a more nuanced understanding of situations by looking into the things that nearly work. For commutativity, we take seriously the fact that $4 + 2$ and $2 + 4$ aren't actually the same process, they just produce the same end result. We look into those processes and find that they can be related by this process of sliding the blocks around each other. But we also notice that there are two different ways of sliding the blocks around each other.

We now have a new question: Do those two different ways of sliding count as the same?

Now things get even more interesting because it really depends if we have another dimension. So far we have decided that we can't slide the blocks around each other in one-dimensional space, but we can do it two ways in two-dimensional space. If we go into three-dimensional space those two ways are related to each other, as we'll see. I am drawn to higher-dimensional mathematics because the questions keep pushing me there when I want to understand more nuance about the answers.

I would now like to give a glimpse of how this tale of commutativity develops in my field of research. We are getting quite far into a piling up of abstraction and so this may seem daunting—feel free to skim over it or just look at the pictures. Conventional wisdom says it's too hard to explain research math to non-mathematicians and so there's no point trying. But I personally think there's still some point trying, even just to give a flavor of how things go and have a chance of piquing interest. It's like the fact that I enjoy looking at the cookbook from the restaurant Alinea, even though the recipes are almost all completely beyond the technical reaches of any normal domestic kitchen. I like the encouraging comments along the lines of "Don't worry if you don't have an anti-griddle, you can just use liquid nitrogen" (because surely we all keep liquid nitrogen in the kitchen). I still like looking at the pictures though, and reading about the lengths that chef Grant Achatz and his team go to in the kitchen. My aim for the next few sections is that, at the very least, you might find it interesting to read about how pure math research goes, even if you don't understand it.

Mathematical braids

In abstract math we study commutativity in general, without having to say whether we're talking about addition or multiplication. We take very seriously the idea of moving objects past each other in space, and we draw a picture of their paths crossing over each other like this picture of A and B moving past each other:

We then start to imagine manipulating these as if they were actual pieces of string, or strands of something. If we had more strands, we could hold two adjacent ones together and bring them across at the same time, but that is deemed to be the same as doing it one at a time. This is expressed in the following equation between braids:

=

We could also bring one strand across two strands held together, and this would be considered the same as doing it one at a time:

=

I hope you can convince yourself that if those were really pieces of string crossing over each other, the equations would represent configurations that aren't really that different in the sense that you could just pull on or nudge the strings in each left-hand configuration to produce the configuration on the right. This is when we count braids as being "the same" in math, if they only differ by nudging and pulling, not by actually undoing something and redoing it.

That is informal and vague, but can be made rigorously logical using some theory of braids that was developed by Emil Artin in the mid-twentieth-century. The theory proves how we can build up braids from "basic building blocks." In this case the basic building block is the single crossing:

It's important that one string is going over the other—if we do the crossing twice we don't get the same thing as doing nothing, even though *A* and *B* end up the same way around as at the start, because the strings get in each other's way:

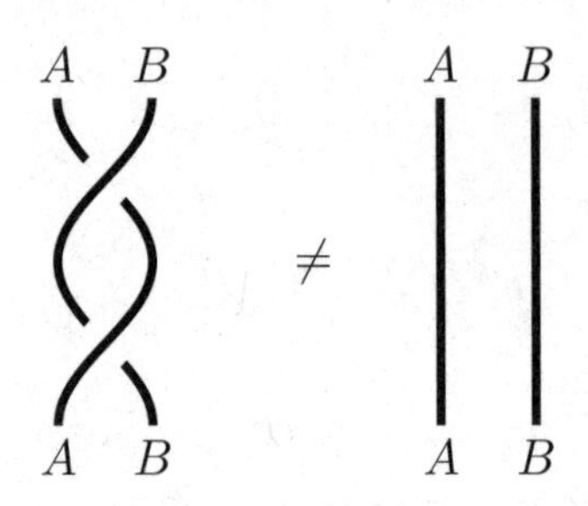

The braids are really showing the *process* by which *A* and *B* are moving around each other.

To actually *undo* the crossing, we need to use the reverse crossing. In the original one, the strand that starts out on the right is on top, but in the reverse crossing the strand that starts out on the left is on top:

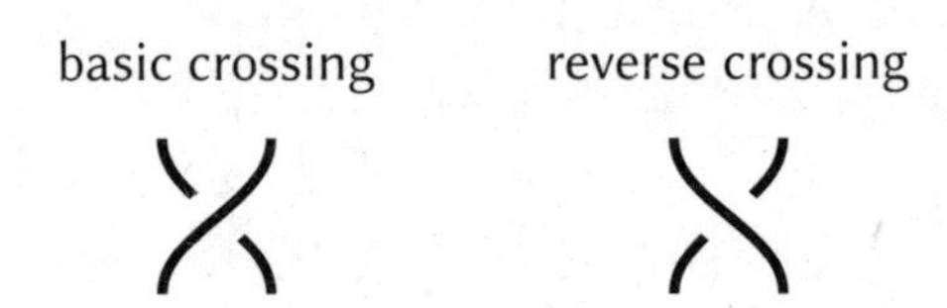

This is the *inverse* of the original crossing, because if you do them in succession, you get something that can be pulled apart. In the world of braids that counts as "the same" as doing nothing.

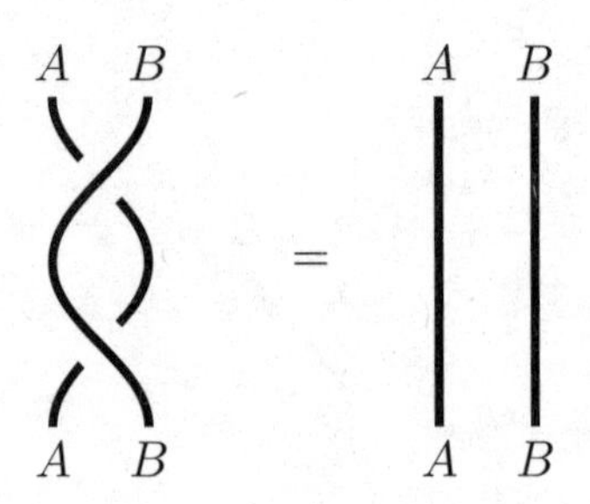

It's a bit like the fact that if you tried to do the braid on the left in your hair, it would just come apart as if you hadn't done anything.

Note that there was an arbitrary choice when we chose the basic crossing, as we could have used the second one as the basic one and the first one as the reverse one. The whole system would end up expressed the other way around but it wouldn't make any difference to what the overall structure is.

Now, for example, we can use just these two crossings (the basic one and the reverse) repeatedly to make a basic three-strand braid as is often done in long hair. We do the basic crossing on the right-hand two strands, and then the reverse crossing on the left-hand two strands, and then keep repeating it. That step-by-step decomposition is shown on the left below, and if we imagine "pulling it tight," the strands smooth out into the braid on the right, which counts as the same braid in math because we didn't untangle anything, we just smoothed it out.

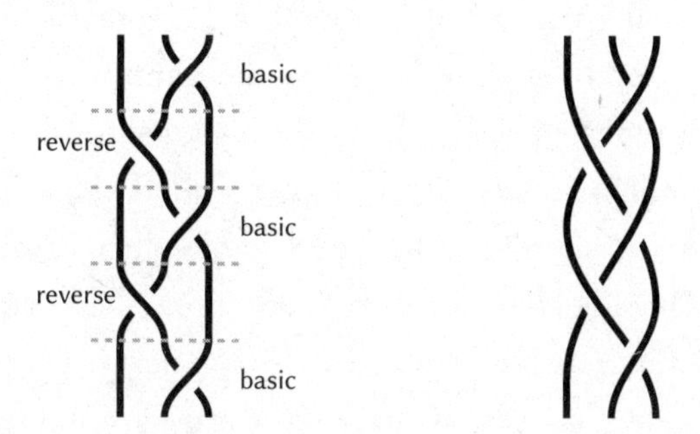

We can think of the basic crossing as being like the number 1, the basic building block for whole numbers. The reverse crossing is like the number −1, which "undoes" the number 1 by addition. We can build all the integers (positive and negative whole numbers) from those, and likewise we can build all braids from the basic and reverse crossings. However, there are more possibilities with braids as we can use any number of strands, so we're not just twisting the same two strands over and over each other.

One situation where we might use more and more strands is in braiding bread. I'm not an expert, but have enjoyed trying to make braided bread:

The braided bread that I am most familiar with is challah, but I recognize that as I am not Jewish I can't truly make challah, as it's not just a bread but a religious ritual. However, bread braiding is an old tradition that is not exclusively Jewish. Swiss folklore says it originated from the old patriarchal custom that if a woman's husband died she was supposed to enclose herself in the tomb with him. At some point this was replaced with the more humane (but still in my opinion sexist and idiotic) practice of her putting a braid of her hair in there, and finally this was replaced with a braid of golden bread, hence the braided Zopf bread of Switzerland. Another possibility is that braided bread was popular as it doesn't go stale as quickly; I suppose this is because the individual strands have more crust around them, but I'm suspicious of this argument, because when I've braided bread the cross-section remains intact as a single cross-section, not as several different strands of bread.

Anyway I find bread braiding mesmerizing and I recommend watching videos of this to see how the patterns are formed from repetitions of simple moves.† I find the movement of the soft dough very

† I like thebreadkitchen.com.

satisfying, and I admit I get extra mesmerized by trying to work out how to translate the techniques into mathematical expressions. The explanations for bread are aiming to make the braiding process replicable by people at home, which is different from the mathematical expressions whose purpose is rigor.

The important thing for the math is to prove that reasoning with the diagrams really is logically sound. In that case we know that we can use our visual intuition to imagine these braids as physical braids. I already mentioned some braids that count as "the same" if we could turn one into the other by just pulling the strings taut a bit. This "sameness" can get much more complicated if we actually pull strings past crossings, or crossings past each other. With the standard hair braid nothing can get pulled past anything else, which is what makes it a secure way of tying your hair back. However, with other configurations of strands we might be able to shift them around to look quite different, without moving the endpoints or having to undo something and re-braid it. In that case the new braid counts as "the same" as the old one.

For example if you stare at these two for a while, maybe you can see that you could just nudge the strands on the left-hand configuration a bit to get to the one on the right of the equals sign, without having to undo or re-braid anything:

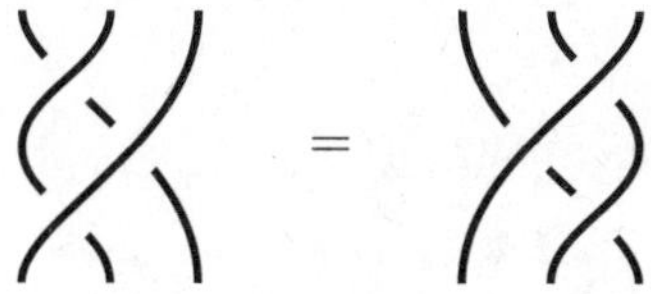

=

One way to see it is to see that in both cases the strand starting at the far right is really on top—nothing goes over it—and it goes down to the bottom left. The strand starting in the middle is next, and it wiggles around a bit but does just end up in the middle at the bottom. Finally the strand that starts at the left is behind everything else and never crosses over anything, and it ends up at the bottom right.

I firmly believe that while staring at that and thinking about it you are doing math, even though you're not trying to calculate an answer to anything, and you're not dealing with numbers.

This means we could take the braid on the left, sort of tug the back strand down, and pull the middle one to the right, and tug the front strand up to the left, and we'd get the braid on the right.

That's the visual intuition but it wasn't at all formal. That's why it's important to prove that the formal and the visual do correspond, and this is an important result in my research field, category theory.

Braids in category theory

When we're thinking about ordinary commutativity of an ordinary multiplication, we just have two choices: either it commutes or it doesn't.

$$a \times b = b \times a \quad \text{or} \quad a \times b \neq b \times a.$$

But if we care about the process of *how* the multiplication can be seen to commute, then we need a more expressive form of algebra. We need a form of relationship that's in between "equal" and "not equal," so that we can measure and record the process of abstractly moving blocks past each other. Category theory is a form of algebra that can do this.

"Category" here is a technical mathematical word, not just an informal word in English. It is referring to a type of mathematical world involving relationships between objects, not just objects. In category theory, those relationships are often drawn as arrows, and are more generally referred to as "morphisms" because they are often a way of morphing one situation into another. So we can now draw commutativity as a process, like this:

$$a \times b \longrightarrow b \times a.$$

However, we often use this symbol $\otimes$ which could represent multiplication, addition, or something else. This is like when we use letters instead of numbers, so that we can talk about numbers in general without having to say what specific number we're talking about. Now we're using a general symbol $\otimes$ instead of a symbol with a specific meaning like + or ×, and we're also still going to use letters to represent the numbers or objects or whatever they are, so we might take two things A and B and put them together as $A \otimes B$. Abstractions do pile up in abstract math.

This abstraction also opens up the possibility of the symbol representing other things, perhaps something a bit like multiplication, even if it is not necessarily actually multiplication. This is like in Chapter 1 when we thought about the possibility of multiplying things other than numbers, and we looked at multiplying shapes together. There are so many things that can be *sort-of* multiplied, that category theory studies the idea of those worlds as an abstract structure in its own right. A set of objects with something like multiplication is called a monoid, and if the sort-of-multiplication commutes then it's called a commutative monoid. A category with a sort-of-multiplication is called a *monoidal category*, as it's a cross between a category and a monoid.

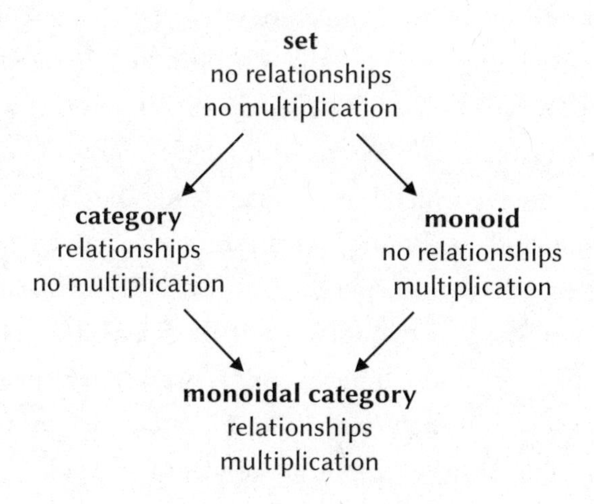

The process of commutativity can then be expressed as a way of "morphing" one situation gradually into another. It's called a braiding in category theory, inspired by the way we braid hair, where the strands of hair record which way around we went. A braiding in a category can be written as a process, or morphism, like this:

$$A \otimes B \longrightarrow B \otimes A$$

representing a process of moving A and B past each other. We ask the braiding to satisfy some basic rules. These rules correspond to the diagrams I drew in the previous section, except that they're now expressed algebraically rather than in pictures. The resulting algebraic structure is called a *braided monoidal category*.

We then prove that if we have any two braid diagrams that count as "the same braid," those will represent the same morphism expressed as algebra in the category. In category theory that's called a *coherence theorem*, as it's how we know that our various different ways of reasoning are coherent with one another.

Coherence theorems fascinate me because they're all about showing that two different ways of reasoning about a structure are equivalent, so that we can make use of both. If one is algebraic and another is visual, then it means we can access different forms of intuition at the same time.

This reminds me of things that are satisfyingly coherent in life. I remember once being delighted that the lid from a jar of olives from one shop fit the jar of olives from another shop; hunting around for lids that match jars (when I'm reusing them) gets quite tedious. I have a little mental note of which lids are interchangeable. My most recent (more random) find was Bonne Maman jam and Claussen pickles.

It might seem that the purpose will always be to invoke our (stronger) visual intuition to help our (weaker) algebraic intuition, and that might well be true in low dimensions. However, as dimensions increase, our visual intuition reaches a limit rather abruptly, and then we might need to rely on algebraic manipulation instead.

Braids in higher dimensions

We can probably go up one more dimension and still imagine things visually. We're still thinking about commutativity, so we're still thinking about sliding blocks around each other. So far we've seen these dimensions:

- In a one-dimensional world, the blocks are stuck on a track and can't move past each other at all.
- In a two-dimensional world, the blocks can move past each other in two different directions.

When we were thinking about whether the two ways of sliding blocks around each other would count as the same, I hinted that it depends on how many dimensions we're in. The key, as usual, is what counts as "the same." If we slide the left-hand blocks around the right-hand ones, does it really matter if we take them the same way around but go a tiny bit further up in the process?

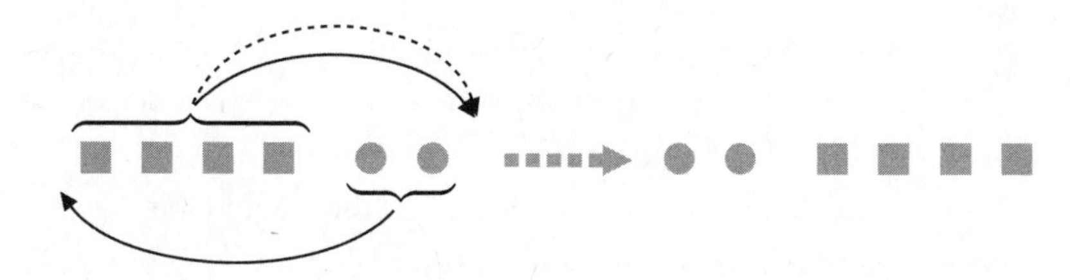

We might not want to count those two ways as really different. In two dimensions, we might well want to consider that there are exactly two ways of doing this slide: one going around the top and one going around the bottom, without worrying exactly where around the top or bottom they go. If we were feeling hilarious, we might want to consider going around and around a few times, but we still probably wouldn't care about the *exact* paths the blocks take to go around and around.

Now what if we're in three dimensions? Then not only could we slide the square blocks a little further up the page, we could also pick them up off the page a little bit. By the same token, that shouldn't really count as an entirely different method. You could then keep edging them further and further off the page, and then down through the air, and eventually you'd get back to the page but *underneath* the circular blocks. So if we have access to three dimensions, going around the top and going around the bottom aren't really that different.

Another way to think about it is that if you think about two humans walking around each other on the ground (two dimensions), we could definitely classify whether they're going clockwise or counterclockwise. Whereas if we think about two birds flying around each other in the sky (three dimensions) it's not really clear what "clockwise" and "counterclockwise" mean anymore, because there's no fixed surface (which is what a clock depends on) and so it would really depend which way we're looking at them.

Mathematically this is saying that we can't really tell the difference between clockwise and counterclockwise in three dimensions, because there is a way of taking the clockwise path and "morphing" it into the counterclockwise path. This is a higher-dimensional version of what we did right at the beginning, which was take 2 + 4 and morph it into 4 + 2, and we found that there were two ways to do it.

Now we've taken the ways to do that morphing, and found that we can in turn morph from one morphing to another. Now things get a little mind-boggling, because there are two ways to do that higher-level morphing: we could either pull the original path toward us (off the page) and down to the bottom, or we could push it backwards (into the page) and then toward the bottom. Perhaps, like with commutativity, we don't care which one we do, we just care that it's possible. Or perhaps (like with braids) we want to record which one we did and measure it. In order to do that, we need yet another dimension in category theory. This takes us to two-dimensional categories, and the higher-level commutativity morphing is called a "syllepsis." Although it's a two-dimensional category, it arises from thinking about paths in three-dimensional space.

The resulting structure is called a *sylleptic monoidal 2-category*. The words pile up on each other along with the concepts, which is definitely something that makes math hard to follow. But we have built up toward these concepts gradually, each time by trying to understand the earlier thing. We start with some objects. We want to understand relationships between them, so we form a category. We want to understand things like multiplication so we have a "monoidal" structure. We want to add subtlety so we have a 2-category. We want to measure what process we used to achieve commutativity, so we get a braiding. We want to measure the way we morph one braid into another so we get a syllepsis.

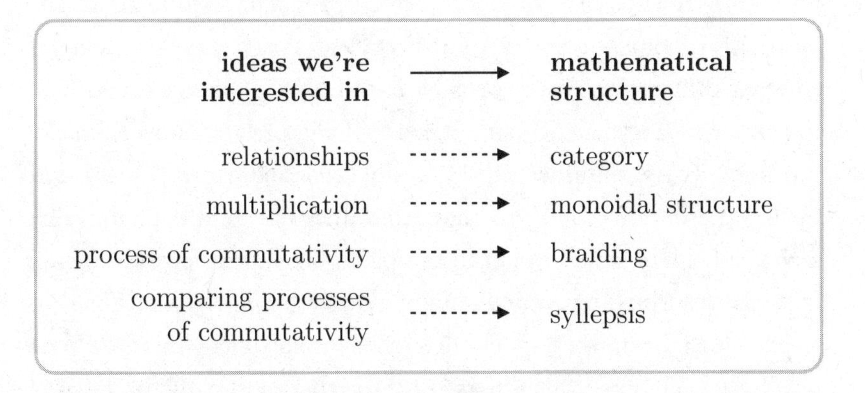

If your brain hasn't combusted yet, you might be able to guess or imagine or predict or infer that this process keeps going. In this many dimensions, the syllepsis around the front and the one around the back are either the same or different, but if we had another dimension we could measure how we get there, and there are two ways around. And if we have another dimension we can measure the difference between those, and so on. What we end up with is a theory of different nuances of commutativity in infinite-dimensional category theory, and this is what a large part of my research is about: how to understand, organize, and analyze those nuances. While our visual intuition can help us a lot in two and three dimensions, it's largely useless in infinite dimensions. We need to fall back on rigorous algebraic methods.

It is far beyond the scope of this book to explain all that in any more detail (and it might have been beyond its scope to explain what I already attempted). If you feel like your brain is exploding then you might have the right idea, because it still makes my brain explode too. It's definitely not "obvious," but at the same time it is something we can get some intuition about by picturing the familiar three-dimensional world around us. I find it endlessly fascinating.

This sort of research even has some direct applications because it's to do with paths in space and how they cross over each other. Even higher-dimensional space is relevant to us despite the fact that we live in a three-dimensional physical world. I've written before about the fact that a robotic arm with a number of hinges is effectively moving in a higher-dimensional space because each hinge requires a coordinate to control it. Our own arm has all sorts of complicated movement determined by the shoulder, elbow, and wrist, each one with back and forth, up and down, left and right, and possibly some rotation. The arm itself sits in three-dimensional space, but its motion is more accurately described by all that data, which might be 8-dimensional or more. So understanding paths in higher-dimensional space is a part of robotics, among other things. And the theory of paths in *abstract* higher-dimensional space is an important part of understanding when complex systems like computers are going to crash. Processes can be described as paths in that abstract space, and crashes can occur when the paths crash into each other, like trying to pull a strand of a braid out when it's firmly locked in.

I don't want to focus on those applications because those are not what drive me, and not what drives pure mathematics. What drives me, and pure mathematics at large, is the wonder and the curiosity and the mystery of it. And all of this came from us thinking of a visual representation of sliding blocks past each other, and then by our continuing to follow our nose and our imagination.

Visual representation of abstract structures

Visual representation of abstract structures is powerful because it enables us to invoke our visual intuition. More than that—it enables us to tap into visual intuition in different ways for the same abstract structure.

My favorite example of this is the standard London Underground map. I can't reproduce it here for copyright reasons, but I hope you can either imagine it or look it up to see what I mean. It is wonderfully designed for clarity for a particular purpose, which is working out how to take the train from one place to another. However, it is not at all geographically accurate. If you look up a geographically accurate version you'll see that it's extremely hard to read. Gazing at the geographically accurate version helps me see how brilliant the standard map is. It's not the original design, but was devised in 1931 by Harry Beck, a technical draftsman who was inspired by drawing electrical circuit diagrams. Those diagrams only focus on connections between parts, not physical placement, and Beck realized that the same could be done for the trains: the important structure to represent in the standard map is only the connections between different lines, not the physical locations of them. So he changed the physical layout of the map while leaving the connections the same. He met with some resistance from the authorities, but a trial printing proved the success among users.

This idea of moving physical placement while keeping connections the same is similar to how we might draw a diagram of factors of 8, in the spirit of the earlier factors of 30. We might start by saying: the factors of 8 are 1, 2, 4, 8. And 8 is 2×4 so perhaps a diagram like this makes sense:

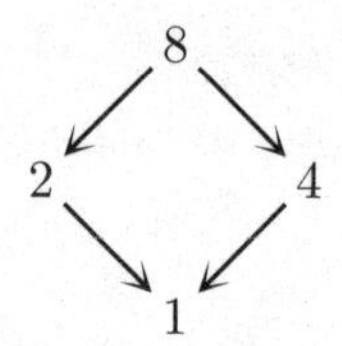

But something is missing: 2 is also a factor of 4 and that's not represented. So we put in an arrow from 4 to 2, and then we find that there is now some redundancy: 8 is like a grandparent of 2 so we don't need the arrow between them, as we can deduce it via 4. Similarly we don't need the arrow from 4 to 1. So now we get this zigzag type diagram:

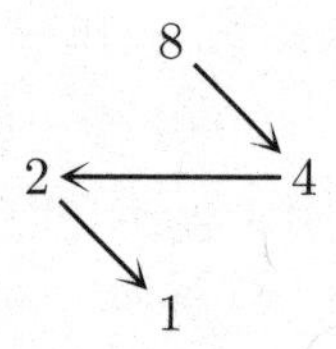

At this point the abstract structure is all in place but it leaves something to be desired physically: it really doesn't need to be in that zigzag. Abstractly the connections all just keep going in a line, so we might as well straighten it out into a line like this:

$$8 \longrightarrow 4 \longrightarrow 2 \longrightarrow 1$$

Like with the London Underground map, we have changed the physical layout without changing the abstract structure. This is a flexibility that is very helpful. Sometimes we might find ourselves drawing a shape that doesn't look very intuitive, like this one:

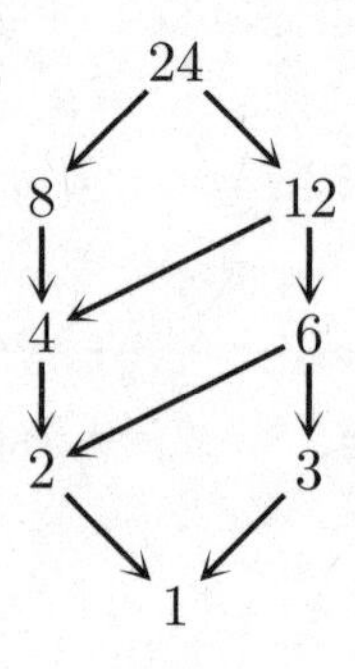

but if we rearrange it a bit we see that it's three squares stacked side by side:

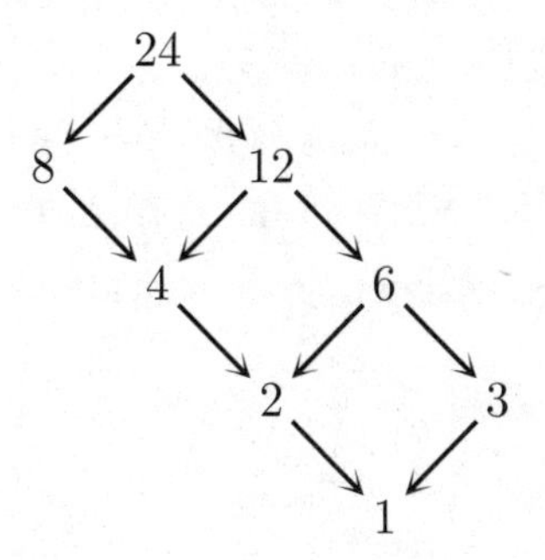

When visual notation retains some physical flexibility it can be a very powerful feature. If we think about how family trees are arranged, the generations have to be ranked physically on the page:

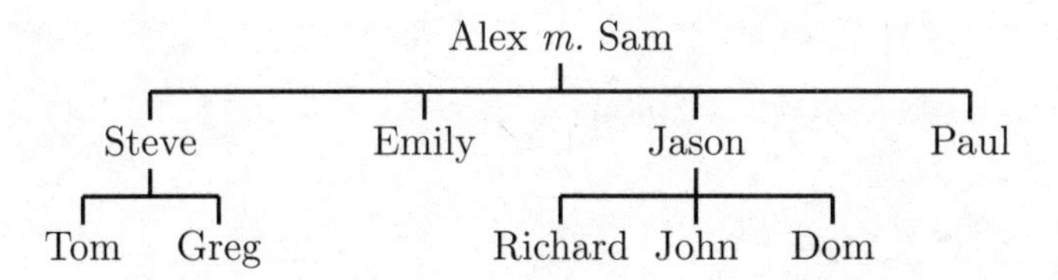

In category theory we use arrows to represent relationships, instead of physical placement on the page. One consequence is that we can change the physical placement without changing the information we're representing. For example we can draw the diagram for the factors of 8 in any of the following ways and it still represents the same abstract information, but the different visual presentations can make different sorts of emotional connections with us.

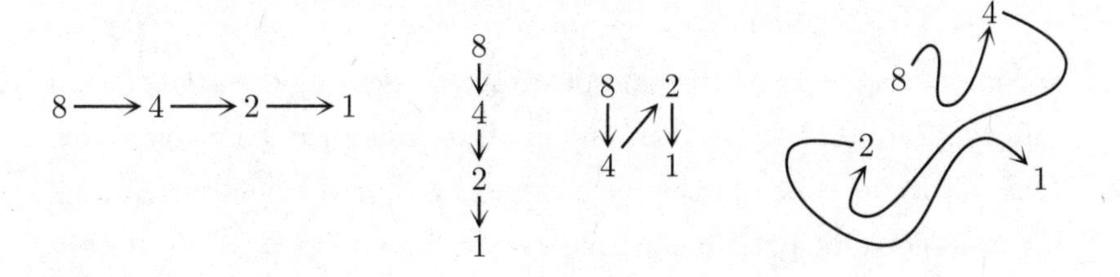

Some of these examples are contrived just to make the point, but the point in question can be used very powerfully to help us reason through situations. Here is a proof from my own work, where these two presentations of the same diagram give very different results visually. For the present purposes it doesn't really matter what these mean, but just from the shape of the pictures I hope you can see that this version doesn't show much, geometrically:

$$\begin{tikzcd} & T^2B \arrow[dr, "\mu^T_B"] \arrow[d, phantom, "="] & \\ T^2A \arrow[ur, "T^2f"] \arrow[r, "\mu^T_A"'] \arrow[d, "Ta"'] \arrow[dr, phantom, "="] & TA \arrow[r, "Tf"'] \arrow[d, "a"] \arrow[dr, phantom, "\Swarrow_\tau"] & TB \arrow[d, "b"] \\ TA \arrow[r, "a"'] & A \arrow[r, "f"'] & B \end{tikzcd} = \begin{tikzcd} T^2A \arrow[r, "T^2f"] \arrow[d, "Ta"'] \arrow[dr, phantom, "T\tau \Swarrow"] & T^2B \arrow[r, "\mu^T_B"] \arrow[d, "Tf"] \arrow[dr, phantom, "="] & TB \arrow[d, "b"] \\ TA \arrow[r, "Tf"] \arrow[dr, "a"'] & TB \arrow[r, "b"] \arrow[d, phantom, "\Swarrow_\tau"] & B \\ & A \arrow[ur, "f"'] & \end{tikzcd}$$

But if we just rearrange it a bit it becomes a cube:

$$\begin{tikzcd} T^2A \arrow[r, "T^2f"] \arrow[d, "Ta"'] \arrow[dr, "\mu^T_A"] & T^B \arrow[dr, "\mu^T_B"] & \\ TA \arrow[ddr, "a"'] & TA \arrow[r, "Tf"] \arrow[dd, "a"] \arrow[ddr, phantom, "\Swarrow_\tau"] & TB \arrow[dd, "b"] \\ & & \\ & A \arrow[r, "f"'] & B \end{tikzcd} = \begin{tikzcd} T^2A \arrow[r, "T^2f"] \arrow[d, "Ta"'] \arrow[dr, phantom, "T\tau \Swarrow"] & T^B \arrow[d, "Tf"] \arrow[dr, "\mu^T_B"] & \\ TA \arrow[r, "Tf"] \arrow[dr, "a"'] & TB \arrow[dr, "b"] \arrow[d, phantom, "\Swarrow_\tau"] & TB \arrow[d, "b"] \\ & A \arrow[r, "f"'] & B \end{tikzcd}$$

This is a powerful idea when we are trying to understand abstract concepts, but unfortunately it's an idea that can be used for nefarious purposes as well, which is what happens when people use misleading visual representation to manipulate unsuspecting readers. It's not strictly wrong because they might still be presenting the same logical information, but they've manipulated it to influence us in some way. One notorious type of example is when a graph of something changing is depicted in a graph like this:

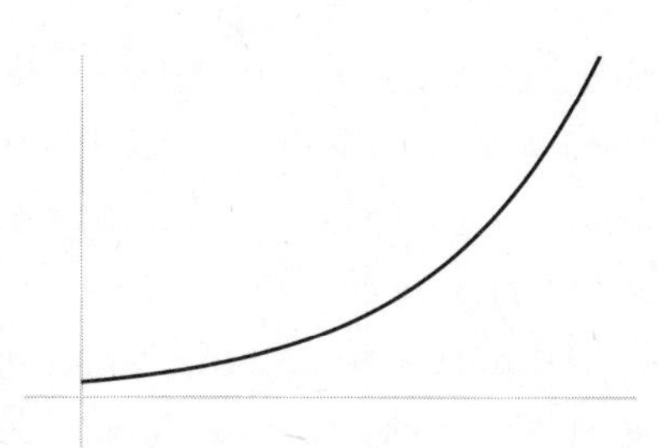

but on closer inspection it turns out that the *y*-axis starts at a million, and that if we started it at 0 the graph would look more like this: not so dramatic after all.

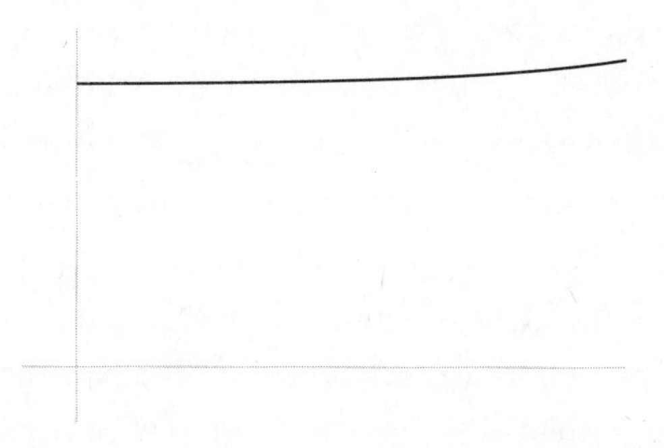

Another example of visual manipulation is a three-dimensional pie chart, where instead of just a circle we see a cylinder. But then we automatically see more of whatever's at the "front," which can make us viscerally get the impression that the "slice" at the front is bigger than it is. This can be used deliberately to make it seem that things in the front are bigger than they are, and things at the back are smaller than they are. In the following example it might be done to detract attention from the amount of water lost to leaks:

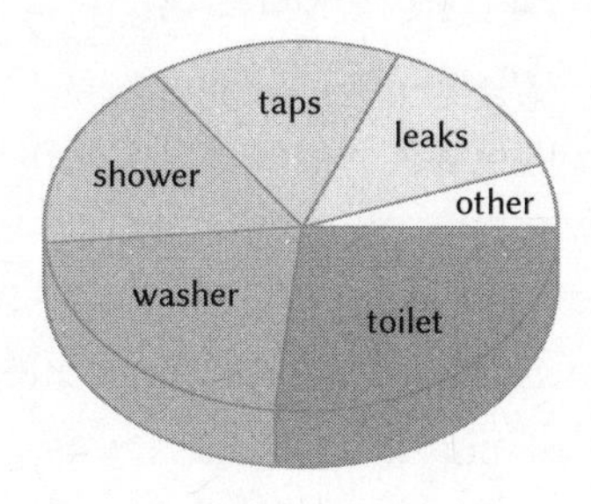

Another case where we are made to think some things are bigger than they are involves some of the most well-known representations of a map of the globe. This takes us back to the relationship between circles and squares, but a dimension up, because we are trying to represent a (more-or-less) spherical earth on a flat sheet of paper. This is doomed to be a distortion, and so we have to make some choices about what we're going to attempt to distort the least. There are many different ways of doing the projection, with different aims. One of the classic ones is the Mercator projection, which is good for navigation as it keeps angles accurate; this means you can accurately work out what direction you need to take to travel somewhere. However, as a result it severely distorts area, and the further something is from the equator, the bigger it looks. This means that northern (imperial) countries look much bigger than they are, and equatorial countries look much smaller by comparison. For example, it makes the US look disproportionately large, and the continent of Africa disproportionately small.

Changing our emotional perception of things without technically changing the content is a clever technique of manipulation. It also happens in language, for good or for ill. If we give something an endearing nickname that can make us feel more fondly toward it, such as the Hairy Ball Theorem, an otherwise arcane theorem which comes to life when described as a hairy ball—it's basically imagining we have a hairy ball and we're trying to comb it neatly. The theorem says we are doomed to have at least one spot that can't be combed smooth. Another one is the Sandwich Lemma, mentioned in Chapter 4, which is actually about understanding a function by squeezing it in between two other ones that we already understand, like some unruly filling in between two flat slices of bread.

This technique of changing our emotional perception can also be used to nefarious manipulative ends, such as when the Affordable Care Act (ACA) was nicknamed "Obamacare" by Republicans who knew that some people who would otherwise support it would viscerally hate it if they were reminded that it was associated with Obama, just because they hate Obama so much. As a result some people declare that they support the ACA but not Obamacare, despite the fact that these are the same thing.

However, if we focus on the underhand ways we can change people's emotional reaction to something, we will lose sight of what a powerful tool it can be for good as well. It is very helpful when looking at infection rates in a pandemic, for example, to use a log scale rather than a linear scale. This means that instead of going up the y-axis in equal spaces, like 10, 20, 30, 40 and so on, we go up by equal *multiples*, like 10, 100, 1000, 10000. This is helpful when we're studying data that is expected to grow by equal multiples rather than by equal increments.

Here is our graph of COVID-19 infection rates from earlier (taking the seven-day running average). On the left is the data plotted on a linear y-axis scale, and on the right is a log scale.

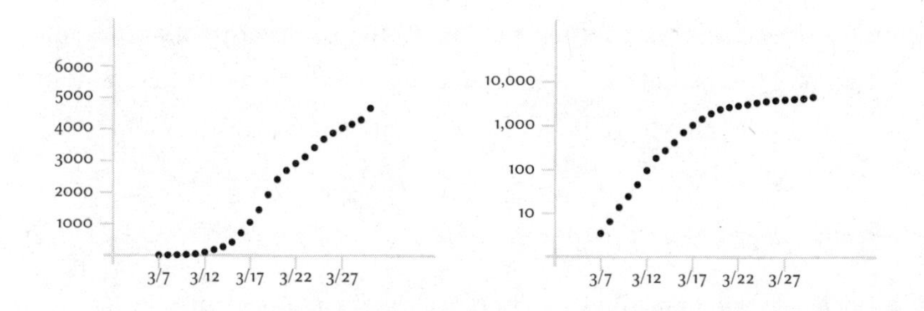

At the beginning, the graph on the right looks like a straight line. This indicates an exponential—an exponential grows by constant multiples, so if we plot it with a y-axis spaced by constant multiples as well, then it will look like a straight line. It is much easier for our eyes to detect a straight line than an exponential curve. It's also then much easier to detect when the rate of growth is slowing, because we can see it "flattening" under where the straight line would have grown to. The key is to be aware that a log scale is being used; a manipulative person could use a log scale without drawing attention to it, to hide how dramatically something is increasing.

This is really the power of turning abstract information into pictures. It's best for us to understand how it works so that we can make sure we are immune to manipulation by unscrupulous people, and also to make sure that we can work out how to communicate our ideas most vividly. I even use graphs to help me understand aspects of my own life.

My life in graphs

I've already shown some of my favorite purely mathematical graphs, and right at the start, in the introduction, I showed a graph of my love of math over time, to contrast my varying love of math *lessons* with my unwavering love of math itself.

Here are some of my favorite graphs that have helped me understand other aspects of my own life. I'm sure it is apparent from all my writing that I like using diagrams to help depict abstract structures to help us understand situations more. Many of those diagrams are ones I have devised to help explain my thinking to other people, but here are some graphs that I have genuinely used to help me understand my own life—and also to go on to explain things to others.

My enjoyment of ice cream

This is a graph of my enjoyment of ice cream against time.

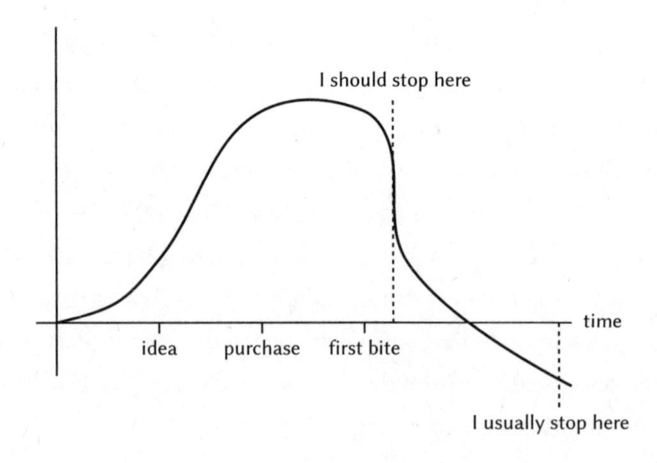

This graph shows that when I start thinking about ice cream I get increasingly excited, and that my excitement peaks more or less when I obtain the ice cream. Even the first bite is not quite as exciting as the act of buying it and the anticipation of it, and then my enjoyment dramatically plummets. Still I usually keep eating it in a futile quest to reattain that lost level of excitement. At some point it actually starts hurting and unfortunately I usually don't stop until I'm actively in pain from eating too much ice cream.

That was how it went in the past anyway. Drawing this graph helped convince me that it would really be better to stop eating the ice cream quite soon after the first bite, when the enjoyment is starting to plummet. That way I can wait until another occasion to get another first bite and peak excitement, and really maximize the enjoyment of the ice cream. Miraculously this means I can now keep a tub of ice cream in the freezer and more or less eat a two-bite serving at a time, and get much more total pleasure out of the same amount of ice cream than when I used to eat a whole tub at once.

Sleep

This next one is a graph of how awake I feel against how much I've slept.

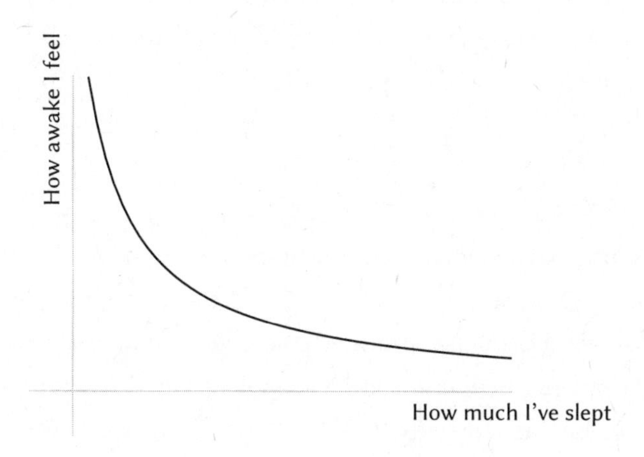

This is a fairly typical graph of inverse proportionality: one thing gets smaller as the other gets bigger. It might seem counterintuitive that I am much more awake when I've slept less, but that's if we're regarding the causation in this direction:

sleep ⟶ awakeness

whereas in reality for me the driving causation is in this direction:

awakeness ⟶ sleep

So of course if I am very awake then I don't need to sleep much, whereas if I am very tired overall then I will need to sleep a lot. This has helped me understand that my level of awakeness is not so much about my sleep last night, but about my general state of life across the most recent several weeks. There are other situations that look paradoxical until you consider the causation differently. For example, if an industry, such as academia, is more accessible to non-wealthy people than to women and people of color, it might seem paradoxical that the industry is more worried about inclusivity toward non-wealthy people than inclusivity toward women and people of color. This doesn't make sense as a response to the current levels of inclusivity, but it makes sense if you consider that people tend to care about their own disadvantages more than those of other people. So if there aren't many women or people of color in an industry then the industry is less likely to care about women and people of color.

People

Some graphs have helped me understand some things about people too. I went to a presentation by Professor Patti Lock which included a section on vivid data visualization. One of her examples came from a study of users on the online dating site OkCupid. The graph showed the ages of people deemed attractive by those seeking dates. There was a plot for women seeking men, and another for men seeking women. The graph looked a bit like this:

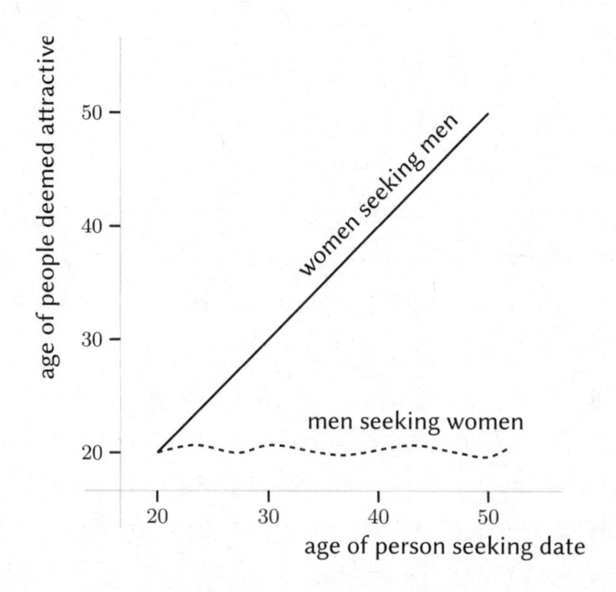

I have just drawn an impression of the graph, but you can see the actual graph on Professor Lock's website.† She delivered these slides with perfect comic timing: she first just showed us the graph for women seeking men, so that we could all think to ourselves "Ah yes, of course, those women tended to find men their own age attractive." Then she revealed the graph showing that straight men on the site find young women the most attractive no matter how old the men are, and the audience collectively rolled its eyes, enjoyed the vivid visuals, and burst out laughing.

How much I care about people

Here is a graph of how much I care about people against how close they are to me (the solid line):

† *Data Analysis in the Mathematics Curriculum*, April 2018, available via www.lock5stat.com/powerpoint.html.

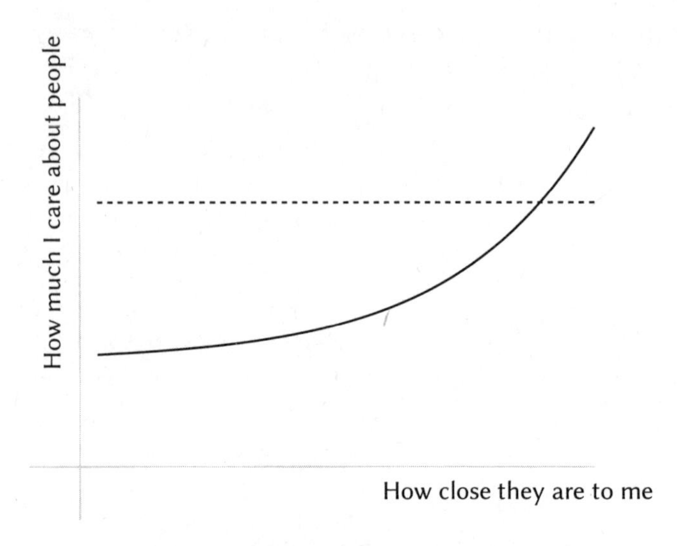

It shows that I care somewhat about everyone, but I care more about people who are close to me. To me, this is reasonable. It's natural to be more upset if something bad happens to a family member than if it happens to a random stranger I've never met. However, there is something contentious about this because some bleeding-heart liberals lament that we care more about people near us than about people on the other side of the world. I do know people who insist on caring about everyone equally; that's the dotted line on the graph. I don't criticize anyone for doing that, but the graph helps me understand that some odd interactions will arise in our friendship.

Heartbreak

Here is a graph that has helped me to understand why heartbreak happens and how we humans could stop doing it. This graph will take a little more narrative explanation, but here it is:

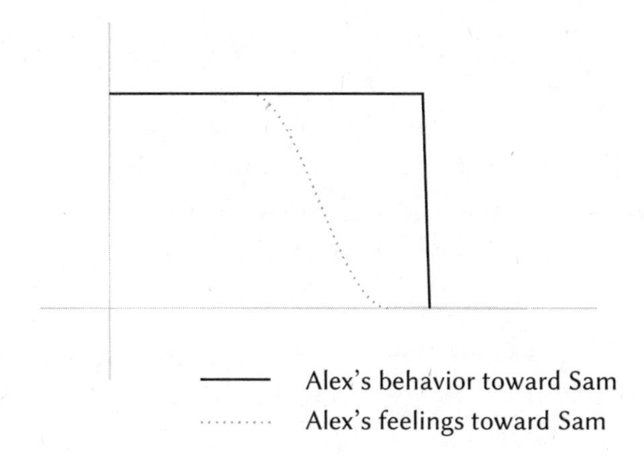

It's inevitable that some relationships come to an end, but almost all cases of heartbreak happen because one person gets dropped off an emotional cliff edge. Yet people rarely grow apart or fall out of love overnight; it happens gradually over a period of time. As I have before, I'll call the people Alex and Sam. In this graph the dotted line shows Alex's gradual decline of affection for Sam. The problem is that while Alex's affection is gradually declining, Alex carries on acting as if everything is fine, or at least Sam continues to believe that everything is fine. Then Alex hits rock bottom and can't take it anymore, and Sam has a rude awakening, suddenly discovering where Alex really is, as opposed to where Sam thought Alex was, and that is where the emotional cliff edge is. That is all shown in the solid, darker line.

Perhaps it turns out that Alex has been having an affair all through that decline, but didn't want to break things off with Sam yet until completely sure and secure in the new relationship. Or perhaps Alex just started feeling a bit unhappy but wasn't sure why, and didn't want to say anything to Sam in case Sam overreacted. So Alex acts as if everything is fine, until suddenly waking up one day unable to take any more.

In either case I really think the key is to make sure that gap between the solid line and the dotted line doesn't happen, so that there is no cliff edge. But that takes self-awareness on the part of Alex, and trust between both parties that they can be honest with each other about what they're feeling. Perhaps if it's not a good relationship then they just can't do that.

So what if Alex gets to the cliff edge and has not managed to be self-aware enough about it? In that case, I really think the best approach is to build a careful slope down. That's the new dashed line shown here:

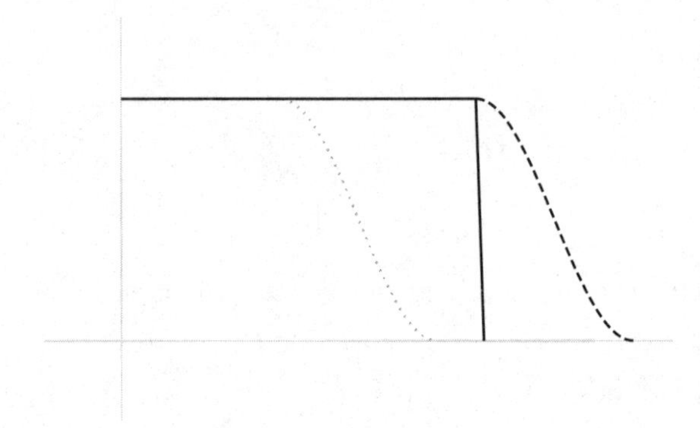

This simulates the original decline, albeit after the event. I have had many people object to this and tell me that it would be dishonest, but my view is that Alex has already been dishonest in hiding the first slope. Moreover, making this new slope would be a form of dishonesty that serves the purpose of protecting another person from injury, and in such cases the dishonesty might be justified. To put it another way, honesty is not a good justification for hurting someone. There are many ways in which we aren't always straightforwardly honest in life, in order to look after someone or be polite or kind. There is a difference between factual honesty and emotional honesty. If someone gives you a gift you don't like, but you say you love it, that is, in a way, factually dishonest, but it's emotionally honest if what you're really saying is that you appreciate their gesture. Similarly if someone says "Do you like my new hair cut?" and you don't, it might be sort of factually dishonest to say you do, but it's emotionally honest if what you really mean is that you wish to be supportive and validating of someone else's choices, which are really nothing to do with you.

To fabricate an emotional slope instead of a cliff edge is, similarly, factually dishonest, but emotionally honest if you are trying not to be an awful person and trying to reduce the amount of heartbreak in the world. Sometimes I think of Shakespeare sonnets or, going even further back, the poems of Catullus, and marvel at the fact that we humans still keep on breaking people's hearts as we have done for thousands of years. And then it feels useless, all this stuff we teach in school, all this progress, traveling into space, building incredibly tall buildings and infinite-dimensional categories, if we are unable to learn how to stop breaking other people's hearts.

It's not something I imagine ever being taught in math at school, but I wish it were, and I do include it at the end of my math course for art students. One student later wrote to me and told me that she told all her friends about this graph, and that just showing them that graph was enough to persuade several of them to take my class.

It's very hard to learn math if you feel no connection to it. For some people the connection comes from getting things right and being praised for it; for some people the manipulation of symbols is fun all by itself; for some people the internal logic is its own satisfaction; for some people the direct applications are exciting. But some people have never been praised in math, feel alienated by symbols and logic, and are left cold by direct applications. They may be more interested by open-ended questions, vivid visual imagery, and indirect applications in the form of light being shed on parts of life that are not typically thought of as connected to mathematics. And in case I haven't already made it clear enough, let me say it again: that is an important part of mathematics. Those things may be dismissed as something "non-math" people say, but they are much closer to the attitudes of professional abstract mathematicians, and we professional abstract mathematicians should spend more time and effort making that known.

CHAPTER 8

STORIES

So far I've talked about how asking apparently naive questions has led mathematicians to develop important new branches of abstract math. Now I want to do something different, and look at how *existing* abstract math addresses some naive questions and spins them into amazing tales scaling great heights, crossing wide oceans, or flying far up into the clouds, starting with a single thought. This is not so much about how we develop new parts of math, but just about how math can take us on these extraordinary journeys. The basic question is just a starting point, like the first clue in a treasure hunt that you think is just going to take you around the room, but then it leads you to the bottom of the garden, and then over the fields and out into the great wild unknown. These are just a few "bedtime stories" told by abstract math, starting from some seemingly naive or trivial questions. Unlike some of the other questions they are not naive questions that turn out to be deep in and of themselves, they are naive questions that spark mathematical stories with surprisingly deep meanings.

How many corners does a star have?

Let's think about a 5-pointed star.

I have always liked these because I can draw them without taking my pen off the page, just by drawing five straight lines directly from one point to the next. Of course, you don't go around the points in order when you do this because that would make a pentagon: you skip one each time. This works because five is odd. We can't make a 6-pointed star like this because if we put six dots around a circle and join them up with lines by skipping one dot at each stage, we'll get back to where we started before hitting all the points, so we'll just make a triangle. That's why to make a 6-pointed star we have to use two overlapping triangles rather than one continuous line, that is, we have to take our pen off the page. Here I've drawn one of the triangles with a dashed line to emphasize the separation of the two triangles.

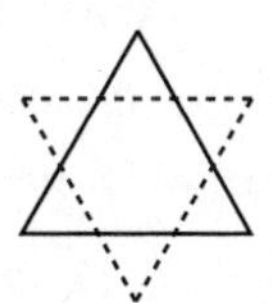

The 6-pointed star is the famous Star of David, a symbol of Judaism.

For a 7-pointed star we can do it like a 5-pointed star again, jumping past one dot at each step, because 7 is an odd number. But now there's another possibility: instead of skipping one point each time we could skip two, which makes a different star. Here are those two different stars—each has seven points, but with different connections and therefore different angles.

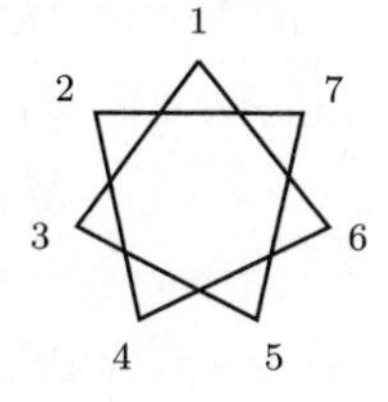

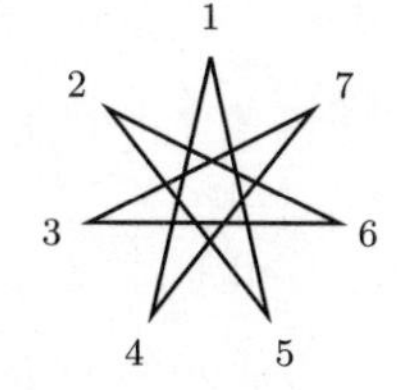

If we try skipping three points at each stage, that's just the same as skipping two but going backwards, so we don't get any different shape. So all in all we get two different ways of drawing a 7-pointed star.

For 8 points, things get more interesting. If we try and make the star by jumping one dot again, we end up with a square, so we can now make the star with two overlapping squares.

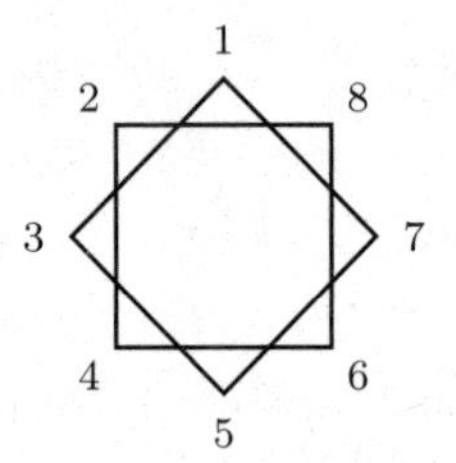

We can generalize that to any even number of points $2n$: we can always make a $2n$-pointed star from two overlapping n-sided polygons.

But we could try jumping two dots. This essentially means going around the points in threes (skip two, land on the third) and this does indeed give us a way of drawing an 8-pointed star without taking our pen off the page.

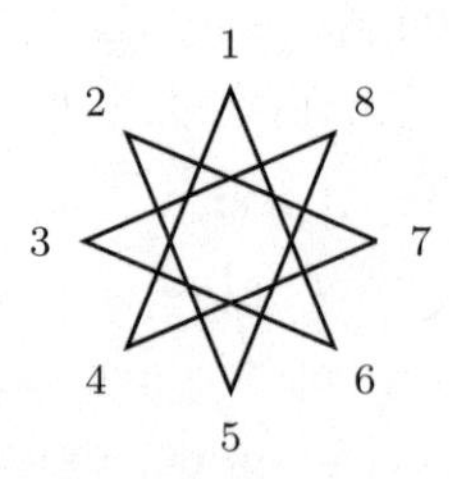

I find this to be a fun way of doodling, but what's really going on is an exploration of the factors of numbers. If you divide the points up evenly then when you jump around you'll come back to where you started too quickly. However, this might also happen if you use a jump that isn't a factor itself but has a common factor with the number of points (other than 1). For example, if you start with 10 points and you jump around in 4s, you won't land right back where you started, but you'll land there after you've hit only 5 points: you'll make a 5-pointed star using just half of your 10 points. So to make a 10-pointed star you could do another 5-pointed star on the remaining points.

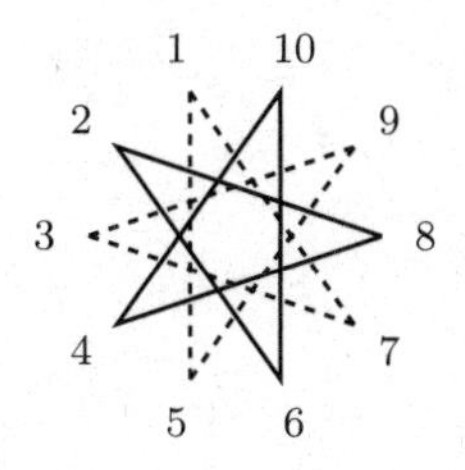

Highest common factors tell us about the patterns we make when we're going around cycles. The highest common factor of two numbers is the largest number that is a factor of both numbers. In the above example the highest common factor of 10 and 4 is 2, but the highest common factor of 10 and 3 is 1. This means that if you jump around 10 points in 3s then you will land on every point before getting back to the beginning.

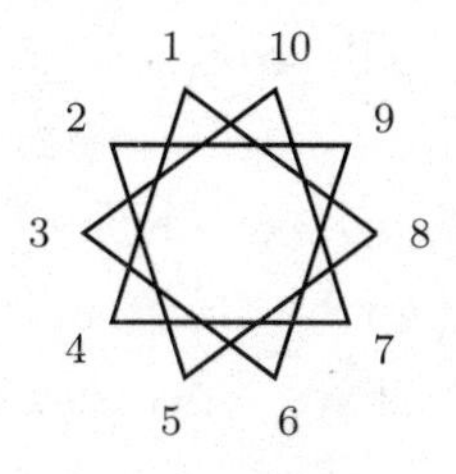

It reminds me of when I lived in Nice and the gym was closed on Sundays. This meant that if I tried to go to the gym every other day I was doomed to try and go on a Sunday every two weeks. But if I tried another pattern like going two days out of three, I was still doomed to need to go on a Sunday. There was no way to do a pattern that avoided Sunday, except by making a pattern across the week rather than across a smaller number of days. This is because 7 is a prime number, so 7 and any smaller number will have a highest common factor of 1.

Doodling with stars has led me into an exploration of highest common factors and cyclic patterns, but it also leads me to think about corners. Going right back to the start, why do we even call this a 5-pointed star? What about the points pointing in? Do they not count as points? What if we talk about corners? How many corners does this star have?

The answer, as usual, is: it depends what you want to count as a corner, and that depends what you're going to do with it. The corners pointing out are arguably more pointy than the corners pointing in, which are more like indentations. You could injure yourself on something pointing out, but you can't really injure yourself on something pointing in (unless you're on the inside of the thing in which case the situation is reversed).

Any time we're counting things in math, whether it's shapes, corners, ways of factorizing numbers, or triangles, we first need to decide what we're going to consider to be one of these things,† and also which of these things we're going to consider to be the same as each other really. So, when counting triangles we first have to decide what "counts" as a triangle (do the edges need to be straight? do they need to be longer than zero?) and then we have to decide which triangles "count" as the same as each other. In abstract math those two steps are by far the most interesting aspect of counting, rather than the actual enumerating part.

For our corners, we don't really have a question of which ones are the same, but we do have a question of which ones count as corners at all. If we want to count the ones pointing in as well as the ones pointing out

† Deborah Stone talks about the impact this has on society in her book *Counting*.

then we might want to say this star has 10 corners. This makes sense especially if we're thinking about the outside shape, because that would be how we count the number of straight lines involved in drawing it.

What if we do want to distinguish between the ones pointing out and the ones pointing in? What is a more rigorous way of talking about corners that "point out" as opposed to corners that "point in"? One way is to think about the angle involved at the corner: if it's less than 180° then the corner will point out, but if it's more than 180° then it will point in.

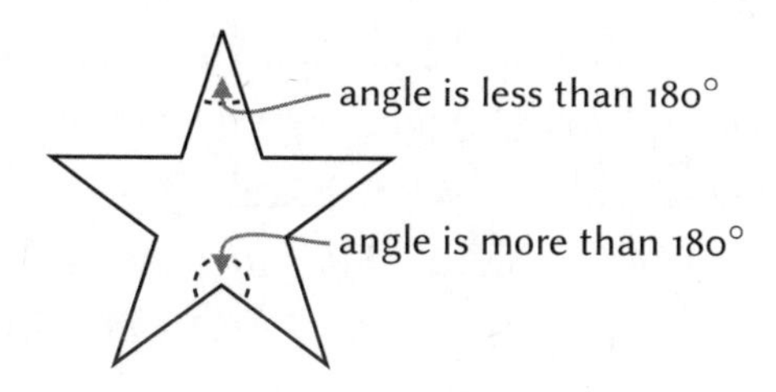

However, in a way that's pinning down something that determines the pointing in or out, rather than a description of pointing in or out. One way that math directly describes pointing in or out is by determining whether you can get from one side to another entirely inside the shape. For a corner pointing out of a star, we can cross straight from one side to the other entirely inside the star, but for a corner pointing in, the thing that makes it point in is that there is a sort of hollowed-out part, which means that to cross straight from one side to another you have to exit the star, to cross over the hollowed-out part.

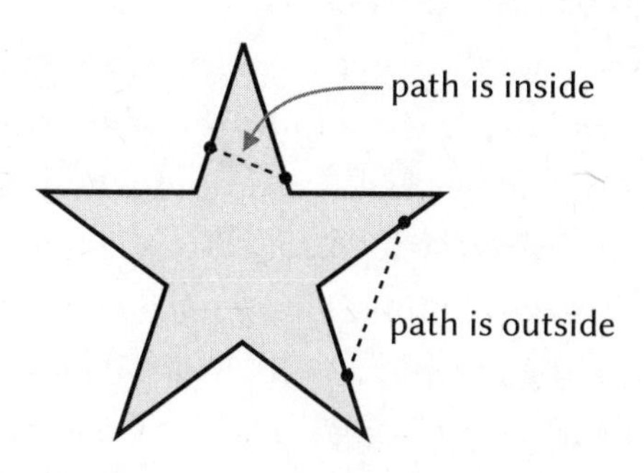

This is the theory of convex and non-convex shapes in math. Convex ones are the ones where you can get from any point to any other in a straight line without exiting. If there are some points where a straight line between them is doomed to leave the shape, then the shape isn't convex. This also relates to how we think about Platonic solids, because the full definition includes the fact that we are looking for *convex* polyhedra; a polyhedron is a three-dimensional shape built from polygons, and polygons are two-dimensional shapes with straight-edged sides. Convex polyhedra are the ones that don't have indentations, and so they are more like spheres, which is why we could eventually use one to build an approximation of a sphere. This is similar to how we can build an approximation of a circle from a polygon with as many sides as we can handle. The more sides we use, the closer it gets to being a circle. This brings us to another question, which I mentioned in Chapter 2, about the sides of a circle.

How many sides does a circle have?

Does a circle have one side, going all the way around? Or does it have no sides, because it has no straight edges? Or infinitely many? Anything in between?

Well, there is a sense in which any of these answers could make sense. It depends what we want to mean by "sides."

I was recently enthralled to discover that one of the famous pizza companies in Chicago makes eight-cornered pizzas. Their standard pizza is rectangular, and it turns out that this is known here as Detroit-style pizza, but I believe there is a rectangular style of pizza in Italy as well. In fact, I had it once when I was young and unaware of these things and I was rather shocked. (And no doubt Italians would be rather shocked by "Detroit-style pizza." If this is translated into Italian, I apologize for making trouble for the Italian translator now.)

Anyway, this particular pizza is known for its crispy, caramelized edges and some people are such a fan of the crispy corners in particular that they want more than four corners, hence the "eight-cornered pizza." When I saw that on the menu my brain went wild and I was agog to know how they did it. Was it octagonal pizza? But then you'd lose some of the crispiness because the angles would be so much bigger. Was it a star-shaped pizza? If so were they only counting the corners pointing out?

I excitedly set about discovering their method, and had completely failed to guess their solution: it's just the same size of pizza but baked in two smaller rectangular halves, making eight corners. I laughed at myself for creating an unnecessarily complicated solution by using inappropriately deep math.

My imagined octagonal pizza holds some critical questions though: it does have more corners, but is it more or less "pointy" than a square? Well, it depends what we mean by pointy. If pointy just means the number of corners, then of course the more corners you get, the more pointy you will be. But then something odd happens because when you have a lot of corners, say, infinitely many, it will pretty much be a circle, which is not pointy at all. This is a hint that maybe that notion of "pointy" is not very stable.

None of that is rigorous (especially the part about infinity) but it is the start of some rigorous ideas of calculus. We've already talked about the polygonal approximation of circles, and in that scheme, we consider a circle to have infinitely many sides, each of which is infinitely short. Well, technically we don't: a circle is still just all the points in the plane that are the same distance away from the center. We can work out the circumference by using the distance around the outside of a polygon with n sides, and then seeing what happens as n approaches infinity. Formally, we define a sequence of numbers where the nth number is the distance around the outside of a regular polygon with n sides. Then we take the limit of that sequence as n tends to infinity. The limit of a sequence of numbers is fully defined in calculus, although the limit of a sequence of shapes is not. So as a way of thinking about circles itself this only works informally, but as a way to calculate the circumference of a circle, it is entirely rigorous.

In fact, this comes down to the general definition of the length of a curve. Calculus builds it all up from the concept of the limit of a sequence of numbers. Then we produce a sequence of approximations of the length of the curve, by drawing straight lines. For example, I can approximate this curve by these straight lines:

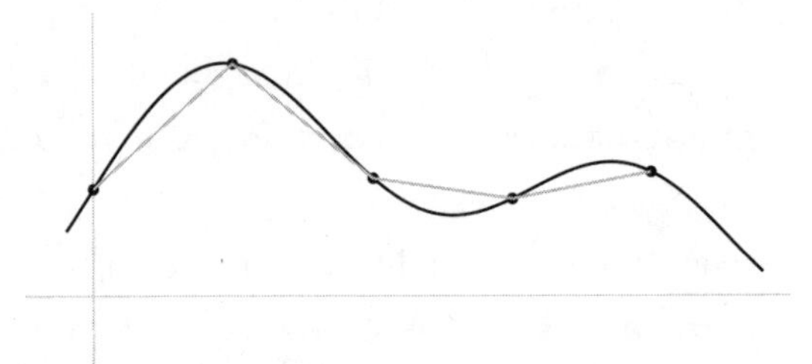

It's not a very good approximation because the dots are so far apart. If I put more dots in, the approximation will get closer. The more dots I put in, the closer it will get. This one with just 9 points is already quite plausible:

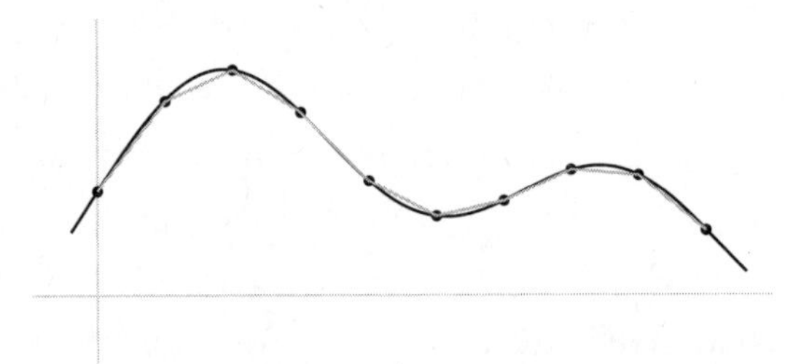

And if I put another dot in between each of the previous ones, it's now fairly indistinguishable to my eyes, anyway.

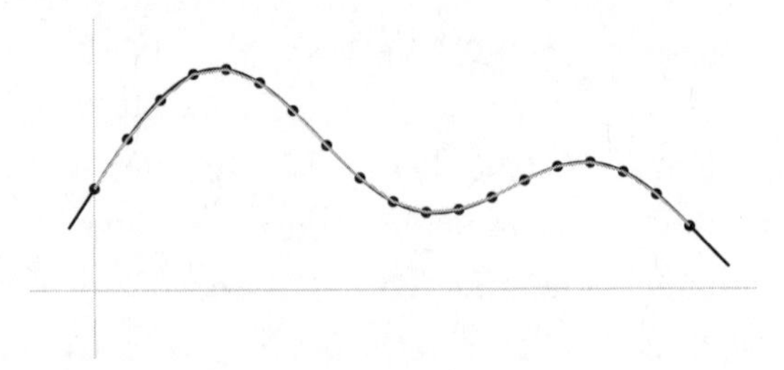

This gives us a sequence of lengths that becomes increasingly accurate: we have a sequence of numbers where the nth number is the length calculated by n equally spaced straight line segments. We can then see what that sequence does as n approaches infinity. That's the definition of the length of a curve.

So perhaps now you're convinced we can think of a circle as having infinitely many sides. But we could also think of it as having one side, if we think of a side as being a path with no corners. And we could think of it as having no sides, because it has no straight edges. But then how many sides does a semicircle have? Does it have one, because it only has one straight side, or two? It feels a bit odd to say it only has one, unless we specify "straight side."

How could we define "side" to give us the (perhaps) more intuitive answer that a semicircle has two sides? We could say that a side is the line (straight or otherwise) between two corners. Then as a semicircle has two corners, it has two sides. This shape has one corner, so it has one side:

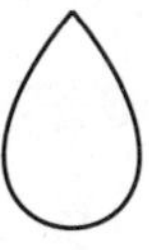

Then a circle is back to having no sides.

In some parts of mathematics, however, we could express a circle as having *any* number of sides, because we can divide it up into as many parts as we want. It's more like the idea that we can "go around in circles" even if they're not perfect circles. In category theory if we had arrows like this

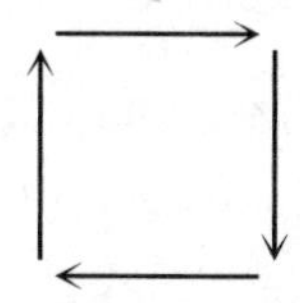

we might say that the arrows are going around in a circle. This is perhaps a little related to the fact that in the field of topology, corners don't matter. It also doesn't matter if a line is straight or curved, or what angles things make with each other—it only really matters how many holes something has. If we only care about holes and we don't care about corners, the number of edges of a shape becomes completely irrelevant and the concept of a triangle just sort of evaporates. In that world, a square is "the same as" a triangle, and so is a pentagon, a hexagon, and a circle. These shapes all count as circles because topology doesn't worry about corners or curvature, only holes. So perhaps a circle can have as many sides as we want.

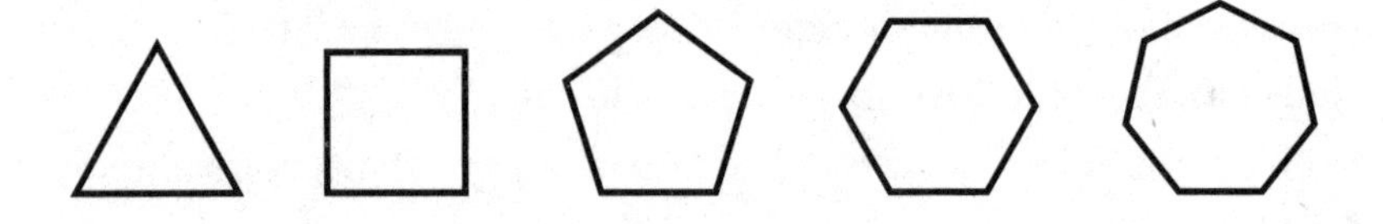

However the shape below is not a circle (I'm considering the white parts to be gaps, so it's like a sort of belt buckle, a circular ring with a bar across the middle) as it has two holes:

This brings me to one of the most contentious questions on the internet, about straws.

How many holes does a straw have?

This is a question that my students periodically ask me, often because another argument about it is raging on the internet. Part of me is heartened that this question captures people's interest, but part of me worries that, like the memes about order of operations, it's just another chance for some people who think they're brilliant to make others feel stupid, by telling them that whatever they think is the answer is stupid.

As usual, what I find mathematically interesting here is not what the answer is, because there are different valid answers depending on what we mean. What is interesting is all the different ways in which we can think about holes, and indeed straws. It goes back to how we count things: what counts as one of the things in the first place, and when two things actually count as the same.

It's valid to say that a straw has one hole, or that it has two, or that it has infinitely many, or even that it has none. That last one might seem surprising, but imagine trying to drink through a straw and finding you couldn't suck anything up through it. You'd probably find it had a hole in it. By that, I mean that it has a hole in it *that it's not supposed to have*, like you might say that your shirt has a hole in it, even though in some other sense it started off with several holes so that you can wear it. As for the straw, you need an intact straw—without a "hole in it."

At the other extreme, we might say that a straw is just made up of a whole load of molecules slightly touching each other, so there are holes everywhere in between them. It's true that's not actually an infinite number of holes, because there will only be a finite number of molecules, but it's a very large number.

Now let's think about the more common argument, which is between those who think a straw has one hole (going all the way through) and those who think it has two (one at each end).

So, what if we closed one end: Would it still have a hole? It would at that point be a bit like a sock. Does a sock have a hole at the top? Personally I think of that as an opening, not a hole, but if we were teaching a child how to put socks on we might well encourage them to pick the sock up and put their foot "through the hole."

As usual with math, I don't think the point is to decide what the right answer is, but to decide in what sense each one is correct. If we think a straw has one hole, how are we defining holes? Likewise if we think it has two holes, we must be defining the holes differently, so what are we doing?

Topology is a branch of math that studies the shapes of things, and it studies which shapes count as the same as other shapes, according to a particular concept of "sameness." It comes from the idea of gradually morphing one thing into another, as if we're squidging a piece of Play-Doh around. Shapes count as the same in topology if you could make one of them out of Play-Doh, and then squidge it around until it becomes the other, without breaking it apart or sticking it together anywhere. This gives rise to the famous idea of a coffee cup being "the same" as a doughnut. The first thing we have to do here is note that the coffee cup has a handle, and the doughnut has a hole.

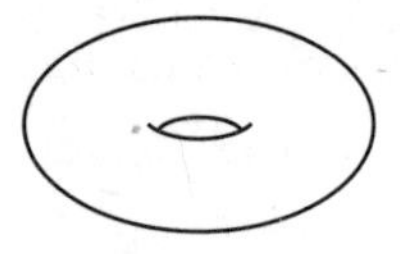

There is a meme that sometimes goes around saying something like this:

> Mathematicians think a doughnut is the same as a coffee cup.
>
> Well try eating a coffee cup for breakfast.

I rather suspect this catches people's attention because they like making fun of mathematicians, possibly specifically mathematicians' unworldliness, or the idea that what they talk about has nothing to do with real life.

The meme somehow presupposes that you know about this form of equivalence in topology, but at the same time that you (possibly willfully) don't understand what the point is. We're not trying to say that a coffee cup is actually the same as a doughnut. Rather, there's a sense in which they're different (that is, in the most obvious contexts of normal life) but also a sense in which they're the same, and sometimes the sense in which they're the same can help us understand something about shapes. The coffee cup example specifically is just illustrative, but the basic idea is that a Play-Doh doughnut shape has a hole going through it, and if you're squidging Play-Doh around there is no way to get rid of that hole without either breaking the Play-Doh (pulling it apart) or filling in the hole (sticking it together). Of course, this concept of Play-Doh is very vague and not mathematically rigorous at all. The mathematically rigorous definition is very technical, but it goes via the concept of a "continuous deformation." This means that we deform the shape so that it becomes another one, but we have to do it in a manner that is "continuous," which means not breaking things apart. (It also means not sticking things together, as that's the same as breaking things apart just with time reversed.)

Anyway, the question then arises: Which shapes count as the same as each other according to this concept of sameness? How many different possible shapes of each dimension are there? A doughnut with one hole turns out to be the same as a coffee cup. But we could also deform the doughnut by flattening it so that it becomes the shape of a pineapple ring or a washer, or, to use the technical term, an annulus.

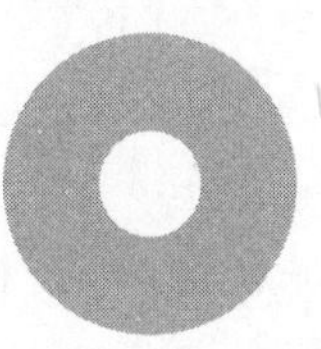

We could instead take our doughnut shape, stick our thumb into the hole and squidge part of the doughnut flat between our thumb and fingers. If we keep doing that working our way around the hole, the doughnut shape becomes a cylinder. We could elongate it so that it becomes a straw. In this sense the straw does indeed count as having only one hole, just like the pineapple ring (annulus). Another way to see that a straw has only one hole in this sense is to imagine making the straw shorter and shorter until it's so short it's basically just a circle, which counts as having one hole.

That's one way that topology deals with holes. Mathematicians don't exactly use the word "hole" because it's rather vague—as we've discovered. And there may be some cases where our intuition really doesn't like what the topology is telling us. We may feel a very strong instinct to call the opening of a sock a hole, even though in topological terms a sock is "the same" as a flat piece of material with an edge but no holes. What about a pair of pants? You might think it has three holes as it has two leg holes and one waist hole. But in topological terms it really only has two holes—if you imagine a two-holed doughnut shape made of Play-Doh, you could turn it into a pair of pants by just elongating it in one direction to make legs.

There is still a very strong instinct to say that the waistband also counts as a hole, and that the opening of a sock counts as a hole. Here is a sense in which this is true: if we zoom in on that particular area of the object then it looks like a surface with a hole cut out of it. If we do this for a straw then we have one hole at each end. What we're really doing is thinking of the straw more as a surface rather than a solid.

This takes us more in the direction of manifold theory. A manifold is a surface that might be very twisted or complicated, but anywhere we zoom in it still looks flat, as long as we zoom in enough. A sphere is an example, because if we zoom in very closely then it looks flat. This is why I still think it's sort of reasonable that people used to think the earth was flat, because they'd only seen a small portion of it. Please note that I don't think it's reasonable nowadays, now that we have such a huge body of evidence that the earth is a sphere (well, a squashed sphere), including photos from space. In order to continue to believe the earth is flat now, you have to deny rather a lot of evidence that most people think is incontrovertible; of course really something can't be incontrovertible if someone contradicts it, because then they have controverted it.

Anyway, the surface of a doughnut is also a manifold, called a *torus.* Now that we're thinking of just the surface (rather than a solid doughnut), it's more like a hollow tube that has been bent around and the ends joined up. And for the sphere we're thinking of a balloon, not a solid ball. The torus has a "hole" through the middle, but it's a different kind of hole—it hasn't been cut out of the surface, it's been sort of built into its shape without breaking the surface anywhere. There are no edges. This is called "genus" in topology. A sphere has genus zero, a torus has genus one, and we can make surfaces of any genus by taking a doughnut with any number of holes we like, and looking at its surface.

Now, if we actually *cut* a hole in a torus, it becomes a "torus with a hole," even though in some other sense the torus already had a hole. The place where we cut one out of the surface is a different kind of hole though, and in math we would call it a "puncture" just like on a bicycle tire, so this new shape has one puncture and also genus one.

We can now think about making a straw by puncturing a surface. We would have to cut two holes in a sphere, and then straighten it up. So in that sense a straw does have two holes. In some mind-boggling way it's like saying that a circle has two holes: the inside and the outside. It's a bit like building a fence around your house and declaring that you've enclosed the whole rest of the world in your fence, and your house is on the outside.

I marginally prefer my definition that involves "locally looking like you have a hole." Math also thinks of it as a circular boundary. Now we're in the realm where a circle doesn't have to be an exact circle, because in topology there's no notion of distance anyway. A hair band counts as a circle no matter how looped up it is. So the opening of a sock is a circular boundary. The opening at each end of a straw is a circular boundary.

So the important thing in this case is to think about how we are defining things, and also to be able to move between the different points of view rather than remain stuck in just one. We have thought of a straw as a predetermined shape, as a collection of molecules, as a solid object, and as a surface. Each point of view gives a different answer for the number of holes; or perhaps each different answer for the number of holes has led us to a different point of view on straws.

In my mathematical view, the most interesting questions in math are the simple ones that are easy to state but have a wide range of possible answers depending on what we want to focus on, what context we're in, what is relevant in those contexts. Math isn't just about numbers and equations. It's about shapes and patterns and ideas and arguments, and in order to reason with those more subtle things we have to make many decisions about how we're going to view them, and how we're going to treat them, which ones we're going to think about now and which ones we're going to think about later. We think about which ones we're going to count as in-some-sense-the-same for now, and perhaps we'll think of some other ones as in-some-sense-the-same later.

It's true that once we've chosen a point of view we then stick to it for a while in order to deduce reliable conclusions about it, just like we can decide on the rules for a game and then play it according to those rules, but if we feel like it we can always go and play a different game or update the rules of this one. It would be wrong to say that math has no rigidity, but also wrong to say that math is entirely rigid. What's crucial, powerful, and beautiful about math is the subtle interplay between the strength of its structures and the flexibility of its points of view. The human body is also rigid and flexible. We have an amazing skeleton that means we can stand upright on a mere two feet, but within that skeleton we have around 200 bones, 360 joints, 600 muscles and 4,000 tendons, meaning that we can move around in innumerable ways. We can run and jump and climb and crawl and sing and smile and dance. That's what math does too. And if we focus only on the rules of logic at its core, we will only notice the rigidity and we will miss the exuberant song and the breathtaking dance that its framework makes possible.

EPILOGUE

Is math real?

As with all the questions we've addressed in this book—and possibly all questions in life—it depends what we mean. What is real? What is realness? Is anything real?

As I said in Chapter 1, most adults don't believe that Santa is "real," but I personally believe that the *concept* of Santa is real, and has a real effect on the world. As an abstract concept, it is real.

This is the level at which math is real. We can't touch it, but there are plenty of other real things that we can't touch either. Sometimes it's for logistical reasons, like the center of the earth or the inside of our own brains. But there are some real things we can't touch because they're abstract, like love, hunger, population density, greed, grief, kindness, joy.

As for almost all questions, I prefer to think about the sense in which different answers are valid. There is a sense in which math isn't real and there is a sense in which it is real. And there's a deeper question lurking, which is: Whether we consider that math is real or not, is that a good or a bad thing?

Math is real in the sense that it's an idea, and ideas are real. They exist. That's good, because if math didn't exist we'd somehow be studying something nonexistent, which makes no sense.

In another sense, math isn't real, if by "real" we are referring to concrete things we can touch, rather than dreams that we create in our heads. But that's also good, overall, although this aspect can make math seem difficult and unapproachable. The power of math comes from the fact that it's not concrete. The abstraction is what enables us to build a strong framework in which logical arguments hold tightly, but the abstraction is also what enables us to maintain flexibility between points of view, unifying a different range of concepts in different contexts, and moving between those different contexts to gain ever more understanding. The abstraction is what enables the structure and also the dance. The abstraction is also what enables us to start with an innocent question and weave so many different stories from it, like a writer who can start with one sentence and then create endless different stories growing out of that tiny beginning. Math is a continual interaction between intuition and rigorous argument, where we use rigor to refine our intuition, and intuition to guide our rigor.

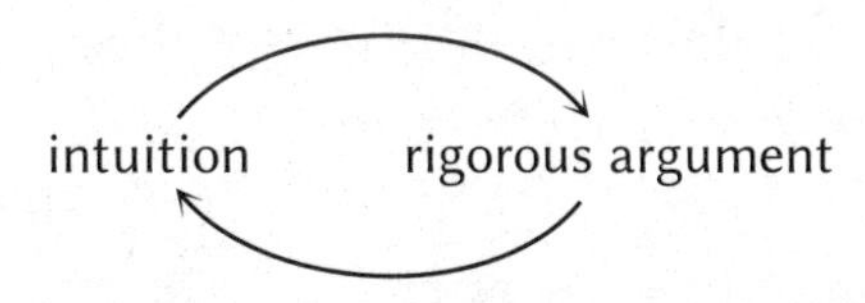

Innocent questions are honest ones that come from curiosity and wonder. These are the best questions. One of the beautiful, powerful, but also mystifying aspects of math is that simple questions can lead to powerful math. This can make it very unnerving to teach, but also very rewarding, because we can start from those questions and go on long and wonderful journeys that give us insight about the real, concrete world.

My hope is that we can all start to enjoy those questions more, whether we are posing them or being asked them, whether we are teachers or students, parents or children, mathematicians or non-mathematicians. I hope that teachers, parents, and mathematicians will all encourage those questions, the innocent ones that are hard to answer, and encourage them especially if you don't know how to answer them. I hope that we can all learn to preserve a child-like approach to math, where we don't expect to understand everything and we don't expect others to understand everything either. Rather, we take every moment of non-understanding as an opportunity to expand our minds, or help someone else's mind expand.

I hope that we will start seeing mathematics as a place to pose questions and explore answers, rather than a place where the answers are fixed and we're supposed to know them. I hope we will also rethink what we celebrate, and place less emphasis on people who get many answers right quickly, and more emphasis on those who are curious, and follow their curiosity on a journey that may be slow and without a clear destination, a quiet walk through the countryside rather than a race to the finish in a sports car.

And crucially, I hope we will give educators more space to bring this kind of math to students at all levels. We should not be calling something education if we are not answering the most innocent and thereby beautiful questions.

One of the most innocent questions is simply about why we are studying any of this in the first place. The temptation is to find concrete applications and "real world" problems for which math can provide specific and precise answers. However, I hope that we will also acknowledge the less specific, but therefore broader purpose of math, which is to help us think more clearly about everything.

If the apparent unrealness of math is off-putting, then one remedy is to present it in a less abstract way. But this has the effect of reducing its power and not showing its true nature, while making it less appealing to those who are more interested in dreams and possibilities than in tools and machinery. A different remedy is to provide a more enticing way into the abstraction, encouraging the dreams, and revealing the power and the possibilities.

If someone doesn't seem interested in reading fiction, it's possible that they will turn out to be more interested in nonfiction, about real people and things. Also it's possible that they just haven't found fiction that appeals to them yet. I love both fiction and nonfiction. Is fiction real? The events might not have taken place in the real world, but the insight into the world is real. I learned much more vivid lessons about how debt escalates from reading *Madame Bovary* than from studying compound interest; I learned more about gender inequality from reading Jane Austen than from studying statistics.

Is math real? The concepts in abstract math might not be part of the concrete world, but the ideas are as real as any other ideas, and, as with fiction, the insights we get into the real world are very real.

And more importantly, whether we consider math to be real or not, it is exciting, mysterious, flexible, awe-inspiring, mind-boggling, satisfying, exhilarating, comforting, beautiful, powerful, and illuminating. Unfortunately there are people trying to keep us out, but they have positioned themselves as gatekeepers at gates that don't need to be there. There are other ways into the glorious dreamworld of abstract math, and I believe we can work together to reveal those scenic paths while also dismantling the gates and removing the obstructions. Then math will be there, quietly waiting, for anyone who wants to go there, anyone who is curious, anyone who is imaginative, anyone who dreams, and anyone who asks questions.

ACKNOWLEDGMENTS

This book was written during a time of unprecedented trauma, globally and personally. I owe thanks to everyone who has helped me keep going in this really awful period of my life. One day perhaps I will write more about this, but pregnancy loss is terrible, traumatic pregnancy loss is worse, and when it causes involuntary childlessness it is a compound trauma that defies description or expression.

Primarily I need to thank my psychologist, Dr. Aisha Kazi, for getting me to a point where I can make it through a day without crying. It is still only a small proportion of my days, but the fact that it is any at all is a huge triumph.

Thank you to Andrew Franklin at Profile and Lara Heimert at Basic for their understanding and continued support.

Thank you to my family.

Thank you to Northwestern Memorial Hospital for saving my life.

Aside from that, I need to thank very many friends. I can't name you all, because every time I try I just start crying. I think you'll understand.

INDEX

Credit: Brian McConkey

Eugenia Cheng is Scientist in Residence at the School of the Art Institute of Chicago and Honorary Visiting Fellow at City, University of London. She has authored numerous titles, including *How to Bake Pi*, *Beyond Infinity*, and $x + y$. Cheng lives in Chicago, Illinois.